Core Concepts for a Course on Materials Chemistry

Core Concepts for a Course on Materials Chemistry

By

T. P. Radhakrishnan

University of Hyderabad, India
Email: tpr@uohyd.ac.in

ROYAL SOCIETY
OF **CHEMISTRY**

Print ISBN: 978-1-83916-669-3
EPUB ISBN: 978-1-83916-715-7

A catalogue record for this book is available from the British Library

The Royal Society of Chemistry is a charity, registered in England and Wales, Number 207890, and a company incorporated in England by Royal Charter (Registered No. RC000524), registered office: Burlington House, Piccadilly, London W1J 0BA, UK, Telephone: +44 (0) 20 7437 8656.

Visit our website at www.rsc.org/books

Printed in the United Kingdom by CPI Group (UK) Ltd, Croydon, CR0 4YY, UK

Preface

The goal of this book is to provide a broad coverage of the core concepts required for a course on materials chemistry at the masters and doctoral degree levels. It is written and organized in a novel style, focusing on a sequential presentation of ideas, concepts, and theories, in a hierarchical manner; the telescoping approach unravels concepts in increasing detail. Graphical illustrations are employed liberally to complement and supplement the analytical arguments and physical models.

The content selection is based on the Chemistry of Materials course that I have been involved in designing, teaching, and updating for three decades. The coverage is intended to provide essential physical and chemical insights into the basic structural and defect aspects of materials, fundamental ideas related to their thermal, electrical, magnetic and optical attributes, and a general introduction to synthesis and fabrication techniques. Since materials science and chemistry encompass ever-evolving themes, this book may be supplemented with discussions of selected materials suited to emerging scenarios; the Postface elaborates further on this point, and the Exercises chapter provides glimpses of possible directions.

I have strived to ensure that the text and graphics are error-free, and would be grateful if any errors noticed while reading this book are brought to my attention. Thank you.

T. P. Radhakrishnan
Hyderabad

Core Concepts for a Course on Materials Chemistry
By T. P. Radhakrishnan
© T. P. Radhakrishnan 2023
Published by the Royal Society of Chemistry, www.rsc.org

Acknowledgements

I am deeply indebted to all the students who have taken the course on Chemistry of Materials that I have taught over the years, interactions with whom have shaped the content and style of this book. The extent of focus on the various concepts throughout the book was tailored by the responses of the students including many of my research students, in the class, in examinations and during informal discussions. My own learning of the subject initiated during my B.Sc. and M.Sc. courses, expanded during doctoral studies under the supervision of Professor Zoltán G. Soos, and the many interactions with the post-doctoral researcher in the group at that time, Dr S. Ramasesha; I owe whatever I have learnt to all my teachers and subsequently my students.

My colleague Professor Ghanashyam Krishna (School of Physics, University of Hyderabad) and Professor S. Ramasesha (Indian Institute of Science) read the complete manuscript and provided highly insightful comments and suggestions; I am extremely grateful to both of them for this invaluable help. I would like to note that the presentations at select areas in this book have benefited from the logical sequences adopted in well-known textbooks like those of C. Kittel, G. I. Epifanov, and H. V. Keer. The final shape of the book is significantly influenced by the guidance provided by Dr. Michelle Carey, Ms. Katie Morrey, Dr. Polly Wilson, and Ms. Amina Headley at the Royal Society of Chemistry; I thank them for the critical support.

I thank my wife Latha and son Vishnu, for their unstinting support throughout the writing of this book, and for sharing their thoughts on several generic as well as technical aspects.

Core Concepts for a Course on Materials Chemistry
By T. P. Radhakrishnan
© T. P. Radhakrishnan 2023
Published by the Royal Society of Chemistry, www.rsc.org

Effective Use of this Book – Note to Teachers

The essential philosophy, motivation, and background that have led to the development of this book are highlighted in the Preface. The contextual placement of the subject of materials chemistry is elaborated in the following Note to Students. I wish to highlight here, the relevance and scope of the book, and how best it can be utilized in teaching.

It has been my experience that, even if a classroom lecture is complete and rigorous with detailed exposition of the relevant concepts, derivations, and illustrations, many students will miss some critical links while listening to the lecture and noting down the salient points. Online teaching modes, in increasing vogue today, may be able to address this issue partly with teaching notes or presentation files made accessible to the students; however, absence of the direct personal touch and the physical ambience of the classroom are always felt keenly. This book is formulated in a specific style. The bullet format and to-the-point presentation are designed to project the concepts and ideas in a systematic order; they are an effective vehicle to collect object groups, selected examples or sequential steps in an analytical derivation. The book is meant to effectively supplement, but not replace classroom lectures; it would support the personal study and augment the learning experience of the student, as a natural follow-up to the classes. For the teacher on the other hand, the directed and point-wise formulation in the book should help to streamline and deliver the classroom lectures effectively and evolve directed assessments.

Core Concepts for a Course on Materials Chemistry
By T. P. Radhakrishnan
© T. P. Radhakrishnan 2023
Published by the Royal Society of Chemistry, www.rsc.org

The model exercises collected at the end of the book are grouped into two types for each chapter, 'content based' and 'context based'. The students may be encouraged to start with the content based exercises and develop new ones of their own; this is primarily a learning process. The context based ones on the other hand can serve as the tools to initiate the understanding process. The model exercises can be used to formulate more, of varying levels of complexity.

An important issue that needs to be addressed in a course on Materials Chemistry is the extent to which specific/specialized classes or families of materials need to be discussed. This point is noted briefly in the Preface of the book, and explained further in the Postface. The enormous variety of advanced materials that are known and continue to emerge makes any selective choice within the time constraints of a course, complicated. However, imparting a flavour of specialized materials and their applications is indeed pertinent. An optimal solution is to first cover the basic conceptual framework provided in this book, and follow it up with brief discussions of selected examples of advanced materials of interest in the emerging scenario and context using appropriate review articles or similar resources. Several of the Exercises provide a beginning in this direction.

Happy teaching!

Effective Use of this Book –
Note to Students

The evolution of human civilization is often described in terms of stages characterized by the dominant materials of each era (Figure 1); the impact of materials on human existence cannot be visualized more vividly. The science and technology of materials encompasses a wide range of physical and natural sciences, and engineering. Chemistry has played a central role in the development and utilization of materials; the nanoscale regime has close analogies to the concept of molecules, the heart and soul of chemistry. An appreciation, understanding, and practice of materials chemistry require a sound foundation in the physical and analytical concepts involved; this book is an effort to serve on this front.

The book has a novel formulation based on a bullet format and to-the-point presentation, designed to project the concepts and ideas in a logical sequence. This should help you to imbibe the key principles that lead to a clear and broad understanding of materials, without having to meander into extensive descriptive text. The liberal use of graphical illustrations is expected to aid this process. An effective way to use the book is to first familiarize yourself with the main topics in each sub-section stated in the initial paragraph(s); they impart a flavour of the principal content. This reading is expected to drive your curiosity, to subsequently glide deeper into the details, derivations or illustrations provided in the bullet steps that follow. The sub-sections are kept compact to facilitate easier assimilation of the essential concepts.

Core Concepts for a Course on Materials Chemistry
By T. P. Radhakrishnan
© T. P. Radhakrishnan 2023
Published by the Royal Society of Chemistry, www.rsc.org

Figure .1   Historic milestones in the evolution of human civilization. Image 1 reproduced from ref. 1, https://doi.org/10.1371/journal.pone. 0160516, under the terms of the CC BY 4.0 license, https:// creativecommons.org/licenses/by/4.0/. Image 2 reproduced from https://commons.wikimedia.org/wiki/File:Copper_tongue_ dagger_(Bellbeaker).png. Image 3 reproduced from https:// commons.wikimedia.org/wiki/File:Bronze_Age_find_in_Slovenia.jpg, under the terms of the CC BY 1.0 license, https://creativecommons. org/publicdomain/zero/1.0/deed.en. Image 4 reproduced from https://commons.wikimedia.org/wiki/File:Iron_Age_Scythian_ Iron_Bells_(28755954795).jpg, under the terms of the CC BY 1.0 license, https://creativecommons.org/publicdomain/zero/1.0/ deed.en. Image 5 reproduced from https://commons. wikimedia.org/wiki/File:Silicon_wafer.jpg. Image 6 reproduced from https://commons.wikimedia.org/wiki/File:Graphene_ SPM.jpg, under the terms of the CC BY 2.0 license, https:// creativecommons.org/licenses/by/2.0/deed.en.

The model exercises collected at the end of the book are grouped into two types in each section, 'content based' and 'context based'. The former essentially requires information already provided in the book, and is meant to guide the process of enquiry. An efficient mode of learning the subject is to take up each sub-section and use the bullet format to create your own questions and answers; the point-wise presentation can guide such efforts. It is often said that, the essence of learning is not so much acquiring the ability to answer questions, but developing the talent to raise relevant questions. The context based problems are natural extensions that follow, and involve utilization of the content of the book in interpretations, analysis, applications or extension scenarios, testing your comprehension. It is strongly recommended that after you go through the exercises, you attempt the development of your own context based ones.

Happy learning!

Reference

1. P. Villa, S. Soriano, R. Grün, F. Marra, S. Nomade, A. Pereira, G. Boschian, L. Pollarolo, F. Fang and J.-J. Bahain, *The Acheulian and Early Middle Paleolithic in Latium (Italy): Stability and Innovation*, 2016, DOI: https://doi.org/10.1371/journal.pone.0160516.

Contents

1 Solid State Structure   **1**

1.1 Types of Solids   1
1.2 Order – Spatial and Dimensional   2
1.3 Symmetry in Crystals   3
     1.3.1 Translation Symmetry   3
     1.3.2 Crystal Systems in 2-D   4
     1.3.3 Crystal Systems in 3-D   5
     1.3.4 Bravais Lattices in 2-D   6
     1.3.5 Bravais Lattices in 3-D   7
     1.3.6 Lattice + Basis = Crystal Structure   11
     1.3.7 Miller Planes   13
1.4 X-ray Diffraction   15
     1.4.1 Interference of Waves   15
     1.4.2 Bragg's Law for Diffraction   16
     1.4.3 Powder X-ray Diffraction   17
     1.4.4 Systematic Absence   20
     1.4.5 Structure Factor   21
     1.4.6 von Laue Condition for X-ray Diffraction   23
     1.4.7 Reciprocal Lattice   25
     1.4.8 Ewald Construction   28
     1.4.9 Structure Factor in Terms of the Reciprocal Lattice Vector   29
     1.4.10 Basic Concepts of X-ray Structure Solution and Refinement   30
1.5 Neutron and Electron Diffraction   31

Core Concepts for a Course on Materials Chemistry
By T. P. Radhakrishnan
© T. P. Radhakrishnan 2023
Published by the Royal Society of Chemistry, www.rsc.org

1.5.1   Neutron Diffraction   31
1.5.2   Electron Diffraction   31
1.6   Common Crystal Structure Motifs   33
1.7   Quasicrystals: A Brief Note   33
References   34

2   Defects and Non-stoichiometry   35

2.1   Point, Line and Plane Defects   35
2.2   Intrinsic Defects   36
2.2.1   Vacancy   37
2.2.2   Self-interstitial   38
2.2.3   Schottky Defect   38
2.2.4   Frenkel Defect   39
2.3   Extrinsic Defects   39
2.3.1   Substitutional/Interstitial Impurity   39
2.3.2   Aliovalent Impurity   40
2.3.3   Charge Compensation Defect   40
2.4   Non-stoichiometry   40
2.4.1   Classifications   41
2.4.2   More Examples of Non-stoichiometric Compounds   43
2.5   Color Centers   45
2.6   Characterization of Defects   45
References   47

3   Thermal Properties   48

3.1   Lattice Vibrations – Phonon Dispersion   48
3.1.1   Phonon Dispersion Curve   49
3.1.2   Lattice With Multi-atom Basis and Vibrations in 3-D   52
3.1.3   Experimental Determination of Phonon Dispersion   53
Curve
3.2   Lattice Heat Capacity   54
3.2.1   Quantization of Lattice Vibrations – Phonons   54
3.2.2   Einstein Model for Heat Capacity   55
3.2.3   Debye Model for Heat Capacity   56
3.3   Thermal Expansion   59
3.3.1   Negative Thermal Expansion   60
3.4   Thermal Conduction   62
References   63

4   Electrical Properties   64

4.1   Free Electron Theory   64
      4.1.1   Free Electron Fermi Gas   64
      4.1.2   Fermi–Dirac Distribution   66
      4.1.3   Electronic Density of States   67
      4.1.4   Heat Capacity of the Free Electron Gas   68
      4.1.5   Electrical Conduction and Ohm's Law   68
      4.1.6   Hall Effect   70
      4.1.7   Thermal Conductivity of Metals   71
4.2   Energy Bands   72
      4.2.1   Energy Spectrum of Electrons in a Crystal   73
      4.2.2   Electron Energy *Versus* Wave Wector: Origin of the   76
              Band Gap
      4.2.3   Bloch Functions   78
      4.2.4   Effective Mass   80
4.3   Metals and Semiconductors   81
      4.3.1   Metals   82
      4.3.2   Intrinsic Semiconductors   83
      4.3.3   Holes   85
      4.3.4   Electrical Conductivity of Intrinsic Semiconductors   87
      4.3.5   Extrinsic Semiconductors   88
      4.3.6   p–n Junctions   90
      4.3.7   Thermoelectric Effects   91
      4.3.8   Hopping Semiconductors   93
      4.3.9   Semiconductor–Metal Transition   95
4.4   Superconductivity   96
      4.4.1   Characteristic Features   96
      4.4.2   Energy Gap   99
      4.4.3   Theoretical Concepts   100
      4.4.4   Superconducting Materials   102
      4.4.5   Superconductor Applications   102
4.5   Dielectrics: Piezo, Pyro and Ferroelectrics   104
      4.5.1   Piezoelectric Materials   105
      4.5.2   Pyroelectric and Ferroelectric Materials   108
      References   110

5   Magnetic Properties   111

5.1   Classification of Magnetic Materials   111
      5.1.1   Paramagnetism and Diamagnetism   111
      5.1.2   Ferro-, Antiferro-, and Ferrimagnetism   113

5.2   Diamagnetism   113
      5.2.1   Larmor Frequency   113
      5.2.2   Langevin Diamagnetic Susceptibility   115
5.3   Paramagnetism   116
      5.3.1   Quantum Theory of Paramagnetism   117
      5.3.2   Pauli Paramagnetism   119
      5.3.3   Adiabatic Demagnetization   121
5.4   Cooperative Phenomena – Ferro-, Antiferro-, and
      Ferrimagnetism   123
      5.4.1   Exchange Energy   123
      5.4.2   Ferromagnetism   125
      5.4.3   Hysteresis   129
      5.4.4   Magnetic Domains   130
      5.4.5   Antiferromagnetism   133
      5.4.6   Ferrimagnetism   134
      5.4.7   Superparamagnetism   136
5.5   Experimental Determination of Magnetic
      Susceptibility   137
      5.5.1   Gouy Method   138
      5.5.2   Faraday Method   139
      5.5.3   Vibrating Sample Magnetometer   140
5.6   Basic Definitions   140
      References   141

6   Optical Properties   142

6.1   Optical Reflectance, Absorption and Scattering   142
      6.1.1   Reflectivity   142
      6.1.2   Dielectric Constant   143
      6.1.3   Excitons   145
      6.1.4   Raman Scattering   146
      6.1.5   Electron Spectroscopy   148
6.2   Photoconduction   148
      6.2.1   Photoresistor   150
      6.2.2   Xerography   150
      6.2.3   Photography   151
6.3   Luminescence   152
      6.3.1   Photoluminescence   153
      6.3.2   Special Mechanisms of Fluorescence Emission   154
      6.3.3   Electroluminescence   156
      6.3.4   Other Forms of Luminescence   157

6.4   Lasers   159
      6.4.1   Population Inversion   160
      6.4.2   Optical Resonator (Cavity)   161
      6.4.3   Diode Laser   162
6.5   Photovoltaic and Photoelectrochemical Effects   164
      6.5.1   Photovoltaic Cells   164
      6.5.2   Photoelectrochemical Cells   165
6.6   Basic Principles of Nonlinear Optics   167
      6.6.1   Linear and Nonlinear Polarization Response   167
      6.6.2   NLO Effects: Second Harmonic Generation   169
      6.6.3   Materials for Quadratic NLO Effects   171
      6.6.4   Electro-optic Effects   173
      6.6.5   Third Harmonic Generation and Organic Polymers   175
      References   176

7   Synthesis and Fabrication   177

7.1   Synthesis   177
      7.1.1   Organic and Inorganic Compounds   177
      7.1.2   Solid State Reactions   179
7.2   Crystallization   184
      7.2.1   Thermodynamic Aspects   185
      7.2.2   Homogeneous Nucleation   186
      7.2.3   Heterogeneous Nucleation   188
      7.2.4   Two-step Nucleation   189
7.3   Crystal Growth from Solution   190
      7.3.1   Phase Diagram   190
      7.3.2   Isothermal Methods   192
      7.3.3   Cooling Methods   193
      7.3.4   Hydrothermal Process   195
      7.3.5   Derivatization Cases   195
7.4   Crystal Growth: Gas, Liquid, Solid States   197
      7.4.1   Sublimation Methods   197
      7.4.2   Melt Growth Techniques   198
      7.4.3   Crystallization in the Solid State   199
      7.4.4   Zone-refining   200
      7.4.5   Sintering   200
7.5   Thin Films   202
      7.5.1   Spin-coating   203
      7.5.2   Thermal Evaporation   204

	7.5.3	Sputtering	205
	7.5.4	Chemical and Electrochemical Methods	206
	7.5.5	Patterning of Thin Films	207
7.6		Colloids and Nanomaterials	208
	7.6.1	Top-down Approaches	211
	7.6.2	Bottom-up Approaches	213
	7.6.3	Some Specialized Nanomaterials/Nanostructures	215
7.7		Monolayers and Ultrathin Films	219
	7.7.1	Self-assembled Monolayer	219
	7.7.2	Layer-by-layer Assembly	220
	7.7.3	Langmuir–Blodgett Film	221
	7.7.4	2-D Materials: Graphene and MXenes	224
		References	225

8 Exercises   226

8.1	Chapter 1	226
8.2	Chapter 2	228
8.3	Chapter 3	230
8.4	Chapter 4	231
8.5	Chapter 5	233
8.6	Chapter 6	235
8.7	Chapter 7	236

Postface   238

Further Reading   240

Subject Index   241

1 Solid State Structure

1.1 Types of Solids

Study of any field is greatly facilitated by a logical and systematic classification of the objects of interest in the study. This is particularly true when we learn about matter and its various characteristics and manifestations. Solids form a central theme in the study of materials and hence understanding their structure is clearly the starting point for learning about them and their attributes and functions.

Classification of solids can take different routes. Some of the general criteria that can be chosen to group solids into different families or classes are the:

- elemental composition of the material
- nature of interactions between the constituent atoms, ions or molecules
- prominent properties of the materials
- functional attributes or applications

The nature of interactions between the building blocks is a commonly used criterion for classifying solids, as it helps in understanding the structure of the material, and the structure–property correlations. Figure 1.1 is a schematic representation of the common types based on such a classification. The general classes of solids are:

- **ionic:** popular examples being NaCl, KCl and MgO
- **covalent:** Si and diamond are the classic systems of this kind

Core Concepts for a Course on Materials Chemistry
By T. P. Radhakrishnan
© T. P. Radhakrishnan 2023
Published by the Royal Society of Chemistry, www.rsc.org

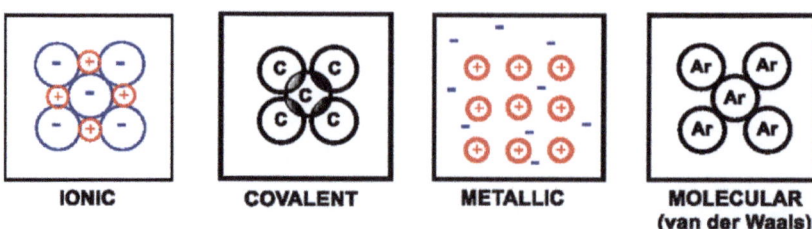

**IONIC   COVALENT   METALLIC   MOLECULAR
                                                                                 (van der Waals)**

Figure 1.1   Classes of solids based on the nature of interactions between the constituent atoms/ions/molecules.

- **metallic:** very familiar in everyday life, Cu and Fe being common examples
- **molecular** (sometimes described also as van der Waals solids): Ar, I_2 and C_{60} are typical members of this family

1.2   Order – Spatial and Dimensional

Order in a solid implies predictability of atom/molecule position and molecule orientation; translational periodicity (see the schematic diagram in Figure 1.2): if moving along a specific direction, an atom/molecule is repeated at a distance r, then it will be found to repeat along that direction at integral multiples of the same distance ($2r$, $3r$, $4r$, . . .).

Even though order is a commonly used term, it can be technically, a relatively complex concept. For example, it can be visualized at different levels:

- **Spatial:** depending on the extent in space (length scale) up to which order (typically, in a 3-dimensional (3-D) structure) persists, the following can be identified
  - crystal (designated also as single crystal, millimeters or higher)
  - microcrystal (micrometers)
  - nanocrystal (nanometers)
  - amorphous solid (none)
- **Dimensional:** depending on the number of dimensions in which, the long-range order persists
  - crystal (typically 3-D)
  - liquid-crystal (with orientational order, typically along 1- or 2-dimensions (1-D, 2-D)); can be further classified into various types such as nematic, smectic, and cholesteric
  - liquid (none, isotropic – implying no long-range order along any dimension)

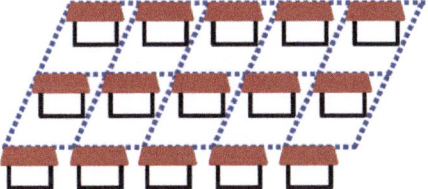

Figure 1.2   Schematic diagram showing translation symmetry in 2-D, along two non-orthogonal directions.

1.3   Symmetry in Crystals

The basic point group symmetry operations are defined based on the condition that at least one point in space is not moved during the operation. These operations, often discussed in the context of molecular symmetry, are:

- identity (E)
- rotation (C_n)
- reflection (σ)
- rotation–reflection (S_n)
- inversion (i)

In addition to the above point group symmetries, crystals with their periodic structure possess **translation symmetry**. In fact, this is a defining characteristic of a crystal. In order to understand the point group and translation symmetries, consider a 2-D square lattice; note that the lattice extends to infinity along the x and y directions. Figure 1.3 shows the lattice; the rotation (by 90°) and translation (by vectors connecting the nearest points, called unit cell vectors) operations leave the lattice invariant (unchanged); therefore these are typical symmetry operations. The operations can be visualized by coloring some lattice points as shown in the figure.

Combination of rotation and translation (along the rotation axis, by a fraction of the unit cell vector) gives a new symmetry operation, '**screw rotation**' in crystals. Similarly, combination of reflection and translation (along an axis parallel to the plane of reflection) gives '**glide reflection**'.

1.3.1   Translation Symmetry

Translation symmetry imposes restrictions on the possible rotation symmetries. This can be geometrically illustrated as shown in

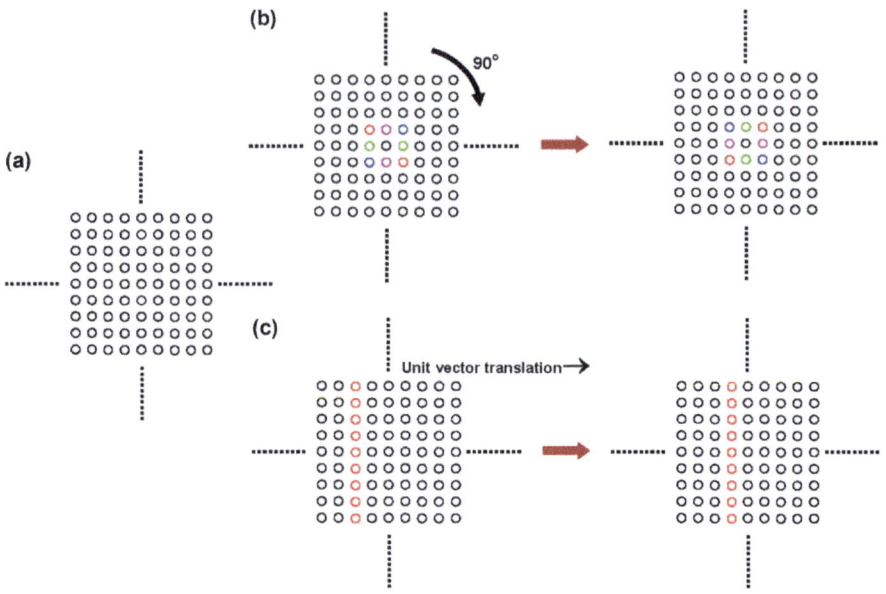

Figure 1.3   (a) 2-D square lattice; (b) C_4 rotation of the lattice; (c) unit vector (along the direction indicated) translation of the lattice.

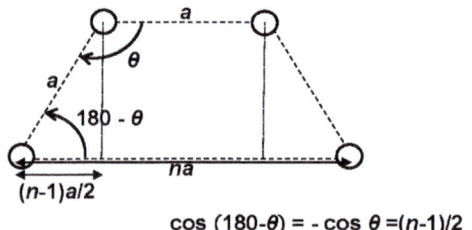

$$\cos(180-\theta) = -\cos\theta = (n-1)/2$$

Figure 1.4   Geometric construction showing lattice points with a unit cell length a along the x-axis, and the relation derived between the angle (θ) of rotational symmetry and the value of n.

Figure 1.4; a is the unit cell length (n has to be an integer, positive, negative or zero, due to the translational symmetry) and θ is the rotation angle allowed by symmetry. Table 1.1 shows that rotations of order 1, 2, 3, 4, and 6 only are compatible with the translational symmetry of the crystal lattice.

1.3.2   Crystal Systems in 2-D

Considering all the possible point group symmetries with the restriction imposed by the translational symmetry, the only crystal systems that can be visualized in 2-D are the following, where a and b

Table 1.1 Possible values of *n* (Figure 1.4) with the condition, $-1 \le \cos\theta \le +1$, the corresponding values of θ and the order of the allowed rotation symmetry operations.

n	3	2	1	0	−1
θ (°)	180	120	90	60	0
Order of rotation	2	3	4	6	1

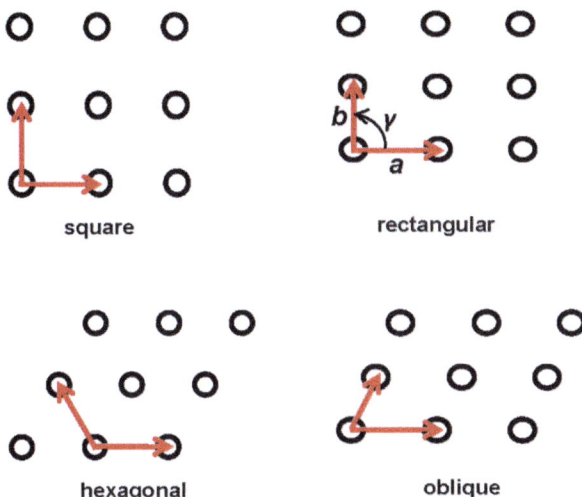

Figure 1.5 Crystal systems in 2-D; unit cell vectors are indicated.

are the unit cell lengths, and γ is the angle between the unit cell axes (shown schematically in Figure 1.5).

- square $(a = b,\ \gamma = 90°)$
- rectangular $(a \neq b,\ \gamma = 90°)$
- hexagonal $(a = b,\ \gamma = 120°)$
- oblique $(a \neq b,\ \gamma \neq 90°)$

It is a useful exercise to list the symmetry elements in each case above. It is also an interesting task to verify and confirm that no periodic lattices in 2-D are possible with a set of point group symmetries that are different from the four cases above.

1.3.3 Crystal Systems in 3-D

Extending the logic used earlier for the 2-D systems, it is found that there are seven crystal systems in 3-D. They are the following, where *a*, *b*, and *c* are the unit cell lengths, and α, β, and γ are the angles

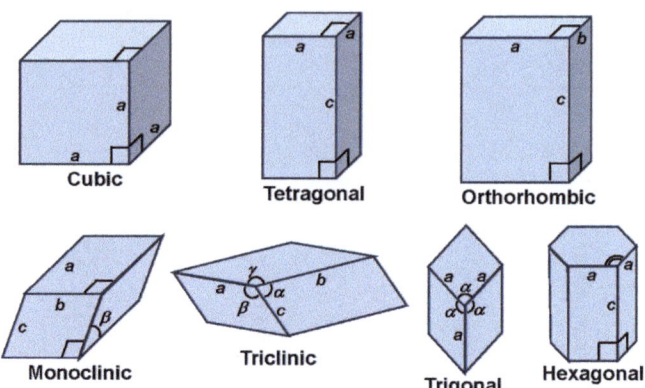

Figure 1.6 Crystal systems in 3-D. Symbol used to mark angles: straight line pairs for 90°, double curved line for 120°, and single curved line for non-90°; unit cell axes are marked by *a*, *b*, and *c*.

between the unit cell axes b and c, a and c, and a and b respectively (Figure 1.6).

- cubic $(a = b = c, \alpha = \beta = \gamma = 90°)$
- tetragonal $(a = b \neq c, \alpha = \beta = \gamma = 90°)$
- orthorhombic $(a \neq b \neq c, \alpha = \beta = \gamma = 90°)$
- monoclinic $(a \neq b \neq c, \alpha = \gamma = 90° \neq \beta)$
- triclinic $(a \neq b \neq c, \alpha \neq \beta \neq \gamma)$
- trigonal $(a = b = c, \alpha = \beta = \gamma < 120°, \neq 90°)$
- hexagonal $(a = b \neq c, \alpha = \beta = 90°, \gamma = 120°)$

1.3.4   Bravais Lattices in 2-D

Adding no further point group symmetries, but incorporating any additional translational symmetry operations possible to the crystal systems, one ends up with five Bravais lattices in 2-D. They are the following (Figure 1.7(a)).

- square
- rectangular
- centered rectangular
- hexagonal
- oblique

It may be noted that the rectangular and centered rectangular lattices have exactly the same set of point group symmetries, but the latter has a new translational symmetry; one way to visualize this

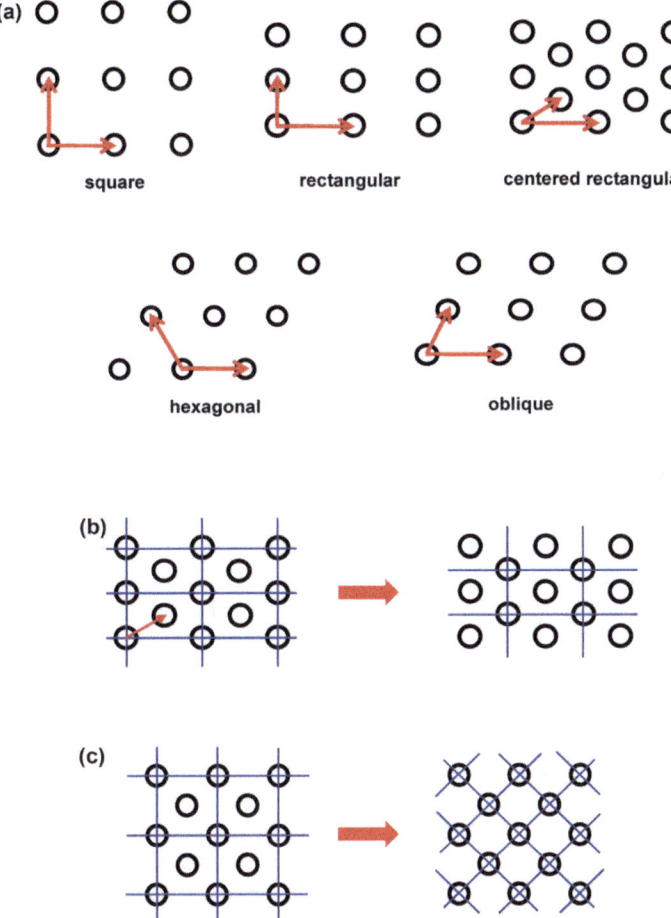

Figure 1.7 (a) Bravais lattices in 2-D. (b) A visualization of the new translation operation available in the centered rectangular lattice. (c) Schematic diagram showing that an imagined centered square lattice is really another simple square lattice with a smaller lattice parameter that is rotated by 45°.

symmetry is to imagine the lattice translation depicted in Figure 1.7(b). A contrasting exercise is to imagine a centered square lattice, and realize that it is just another simple lattice (Figure 1.7(c)), and therefore not a new Bravais lattice.

1.3.5 Bravais Lattices in 3-D

Similar to the case of the Bravais lattices in 2-D, these follow from the crystal systems shown in Figure 1.6. The 14 Bravais lattices are listed

Table 1.2 Bravais lattices in 3-D.

Crystal system	Bravais lattices	Number
Cubic	P, F, I	3
Tetragonal	P, I	2
Orthorhombic	P, C, F, I	4
Monoclinic	P, C	2
Triclinic	P	1
Trigonal	P	1
Hexagonal	P	1
Total		**14**

in Table 1.2, with the standard notations, **P** (primitive), **F** (face-centered), **I** (body-centered), and **C** (edge-centered).

As an example, consider the three Bravais lattices under the cubic crystal system (Figure 1.8(a)); they all belong to the same point group, O_h and hence have the same set of point group symmetries. However, they possess different sets of translational symmetries; for example, the bcc lattice has a characteristic half body-diagonal translation.

It is an interesting exercise to understand why the face-centered and body-centered lattices are the same in the tetragonal system, but not in the cubic system. This has to do with the fact that $a = b$ in tetragonal, and the F and I lattices are the same with only a rotation of the a, b axes system, and change of magnitude of the parameter a (Figure 1.8(b)). The orthorhombic system has a new lattice, C; if a similar situation is imagined in the tetragonal system, it would simply be another tetragonal P lattice with a different $a = b$ value (extension of the idea illustrated in Figure 1.7(c)).

The hierarchy of crystal symmetries at this stage can be summarized as shown in Table 1.3.

Before proceeding further on the symmetry in crystals, a little more insight into the concept of a **Bravais lattice** is useful. In 3-D, a Bravais lattice can be imagined as arising from the collection of all points represented by the set of vectors,

$$\vec{R} = n_1 \vec{a}_1 + n_2 \vec{a}_2 + n_3 \vec{a}_3$$

where n_1, n_2, n_3 are integers and $\vec{a}_1$, $\vec{a}_2$, $\vec{a}_3$ are primitive vectors that determine the symmetry of the lattice.

- Primitive vectors for a cubic lattice are,
  $\vec{a}_1 = a\hat{x}$; $\vec{a}_2 = a\hat{y}$; $\vec{a}_3 = a\hat{z}$ where a is the unit cell length and $\hat{x}, \hat{y}, \hat{z}$ are the unit vectors along the three Cartesian axes.

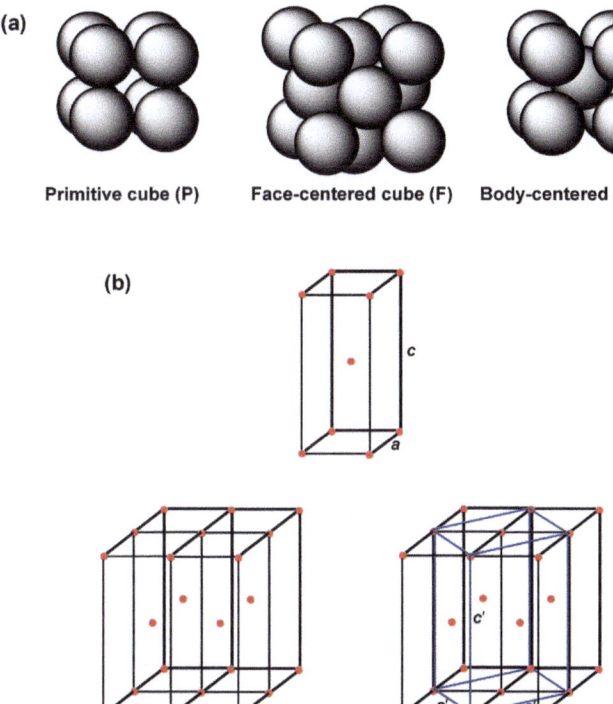

Figure 1.8 (a) Bravais lattices in the cubic system. (b) Unit cell of a body-centered tetragonal lattice (red filled circles indicate the lattice points and the cell parameters *a*, *c* are shown), and a collection of four adjacent unit cells showing that the body-centered lattice can also be viewed as a face-centered lattice (blue outline with $a' = \sqrt{2}\, a$, $c' = c$).

Table 1.3 Hierarchy of crystal symmetries with spherical basis.

Point group operations	7 Crystal systems
Point group operation + translation symmetries	14 Bravais lattices

- Similarly, the primitive vectors for the face-centered cubic (fcc) and body-centered cubic (bcc) lattices are (these are convenient, but not unique choices):
  - fcc: $\vec{a}_1 = \dfrac{a}{2}(\hat{y} + \hat{z});\ \vec{a}_2 = \dfrac{a}{2}(\hat{z} + \hat{x});\ \vec{a}_3 = \dfrac{a}{2}(\hat{x} + \hat{y})$
  - bcc: $\vec{a}_1 = \dfrac{a}{2}(\hat{y} + \hat{z} - \hat{x});\ \vec{a}_2 = \dfrac{a}{2}(\hat{z} + \hat{x} - \hat{y});\ \vec{a}_3 = \dfrac{a}{2}(\hat{x} + \hat{y} - \hat{z}).$
- The primitive vectors define a primitive unit cell; Figure 1.9a shows the fcc cell.
- Primitive vectors for the tetragonal and orthorhombic lattices are:
  - tetragonal: $\vec{a}_1 = a\hat{x};\ \vec{a}_2 = a\hat{y};\ \vec{a}_3 = c\hat{z}$
  - orthorhombic: $\vec{a}_1 = a\hat{x};\ \vec{a}_2 = b\hat{y};\ \vec{a}_3 = c\hat{z}.$

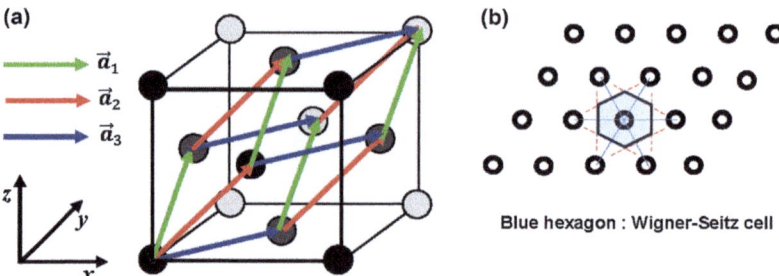

Figure 1.9 (a) Primitive vectors and primitive unit cell of an fcc lattice (the lattice points at the front, middle, and back are shown in black circles with black, dark grey and light grey filling respectively); (b) construction of the Wigner–Seitz cell for a 2-D hexagonal lattice.

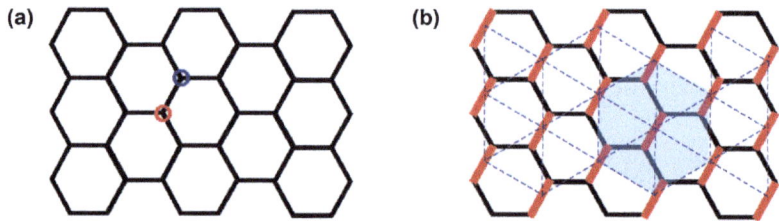

Figure 1.10 (a) Honeycomb lattice is not a Bravais lattice; notice the differences in the points marked with blue and red circles. (b) A combination of these points forms the hexagonal 2-D Bravais lattice shown by the broken blue lines; the unit cell is also indicated.

As mentioned above, the unit cell can be chosen in multiple ways. A special choice is the **Wigner–Seitz cell**, constituted by all points that are closer to one lattice point than any other; it is the innermost region enclosed by perpendicular planes (lines in 2-D) bisecting the lines connecting a lattice point to all the neighbors (Figure 1.9(b)).

Every lattice point in a Bravais lattice has identical environment; observation of the surrounding, from any point is identical. This aspect can be used to distinguish a non-Bravais lattice from a Bravais lattice.

- For example, a honeycomb lattice (Figure 1.10) is not a Bravais lattice, as the points ○ (red) and ○ (blue) do not have identical environments; however, a combination of two of these points leads to the hexagonal Bravais lattice in 2-D.
- A cubic close-packed (ccp or fcc) lattice is a Bravais lattice, but a hexagonal close-packed (hcp) lattice is not; Figure 1.11 illustrates this interesting point.

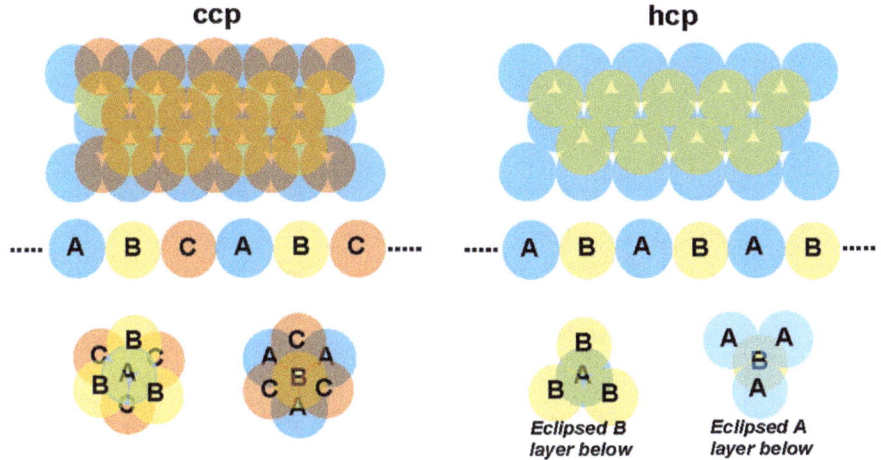

ccp **hcp**

····· A B C A B C ····· ····· A B A B A B ·····

Eclipsed B layer below *Eclipsed A layer below*

A and B have identical environments   A and B have different environments

Figure 1.11   Cubic and hexagonal close packing; the layer sequences are shown for each. The identical environment of each site in ccp and the different environments in hcp are shown schematically.

1.3.6   Lattice + Basis = Crystal Structure

The discussion so far has looked at the symmetries of the lattice; '**lattice**' is simply a set of points in space described by a set of coordinates, two in 2-D or three in 3-D. A real crystal is made up of atoms, ions, or molecules; the molecule can be as small as H_2 or a large protein.

- The object or set of objects (with atoms at specific locations with respect to each other) placed on the lattice points, is described technically as the '**basis**'.
- A crystal consists of the **basis** organized on a **lattice** with a specified symmetry.

The basis can be spherical (perfectly symmetric), *e.g.*, a single atom, or non-spherical (with lower symmetry than a sphere) if it has more than one atom, a molecule, *etc.*

- Symmetry of the unit cell remains unchanged when a spherical basis is added.
- However, symmetry of the unit cell is reduced when a non-spherical basis is added; it goes to one of the sub-groups of the original point group.
- The example below, of the square lattice unit cell (Figure 1.12(a)) illustrates the point.

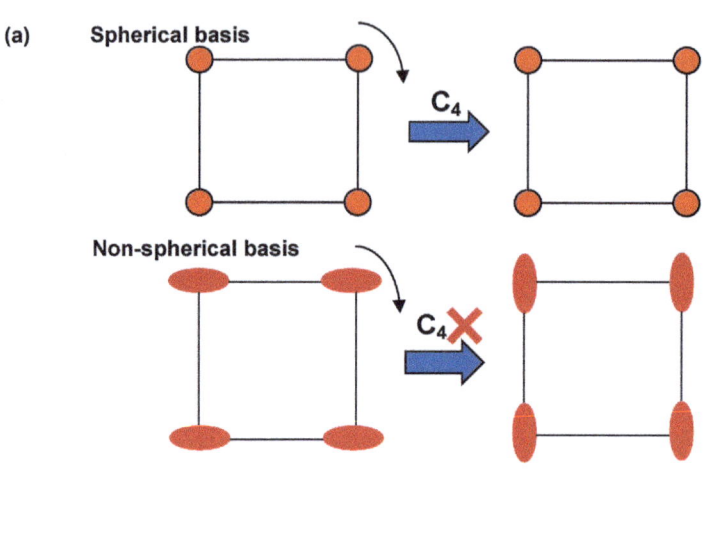

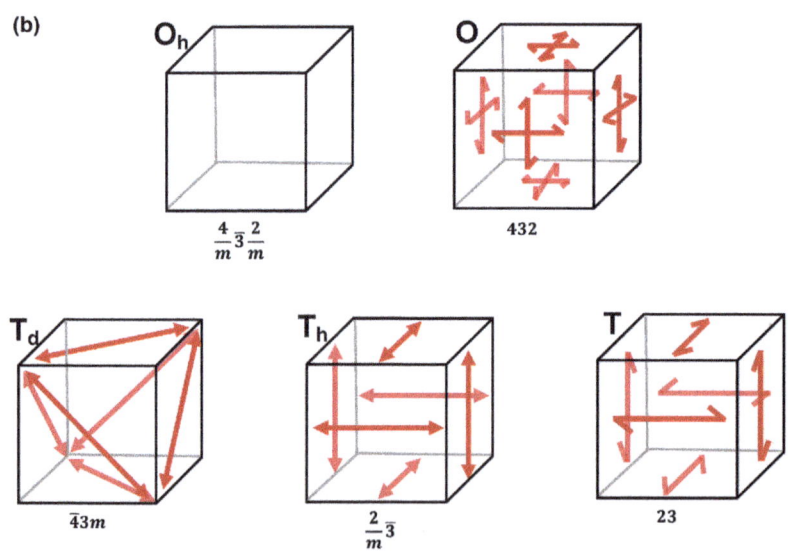

Figure 1.12 (a) C_4 rotation operation shows that the square lattice with a spherical basis possesses C_4 symmetry, whereas with a non-spherical basis does not. (b) Schematic diagrams showing the symmetry of the five cubic point groups. The point group of each is indicated at the left top and the international notation is shown at bottom; **2**, **3**, **4** are rotation axes, **3̄**, **4̄**, are improper rotation axes, and $\dfrac{2}{m}, \dfrac{4}{m}$ are rotation axes with perpendicular mirror planes.

Table 1.4 Hierarchy of crystal symmetries.

	Lattice + spherical basis	Lattice + non-spherical basis
Point group operations	7 crystal systems	32 crystallographic point groups
Point group operations + translation symmetries	14 Bravais lattices	230 space groups

In a similar way, subgroups of the various crystal systems in 3-D (Figure 1.6) can be identified. The different point groups based on the cubic system are shown in Figure 1.12(b).

The final hierarchy of crystal symmetries, building on the classification shown earlier in Table 1.3, can be summarized as shown in Table 1.4.

1.3.7 Miller Planes

Miller indices are a convenient way of designating planes in the crystal lattice; this will have a direct bearing on the discussion and analysis of the X-ray diffraction process later. We look first, at 2-D lattices in which they should strictly be called Miller lines. The examples in Figure 1.13 show the designations and the distance between the lines. Note that the distances decrease with increasing values of the indices.

The logical steps involved in naming the Miller planes can be explained using the example shown in Figure 1.13(c).

- Consider the point at an arbitrary origin (0, 0) in the x, y axes framework.
- The 'plane' intercepts the axes at fractional coordinates 1/2 and 1/3 respectively.
- Take the inverse values; if not integers, scale them to the lowest set of integers: 2, 3.

The distance between the planes can be calculated using simple geometry. In the case of the square lattice, following the above protocol, a 'plane' $(h\ k)$ will intercept the axes at a/h and a/k. The distance between the planes, d, can be obtained using the geometry of the right-angled triangle as shown in Figure 1.13(e).

For a cubic lattice (3-D), the lines in Figure 1.13(c) would extend parallel to the z axis to form planes. Correspondingly,

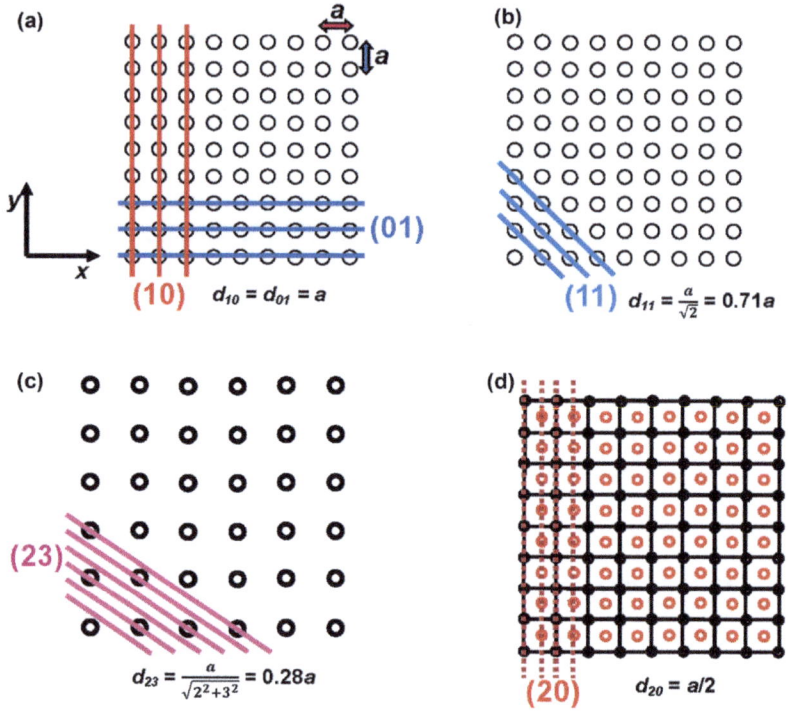

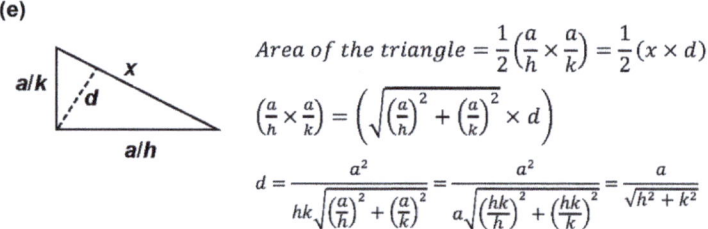

Figure 1.13   Miller 'planes' in a 2-D square lattice, with unit cell vectors along the x and y axes and length a. 'Planes' (a) (1 0) and (0 1), (b) (1 1), (c) (2 3), and (d) (2 0). The distance between the 'planes', d, in each case is shown. (e) A geometric derivation of the equation for d in a 2-D square lattice.

- the intercepts will be $a/2$, $a/3$ and ∞
- the Miller plane will be designated as (2 3 0)

Some common Miller planes are shown in Figure 1.14. Based on the above discussion, the interplanar distance, d for an $(h\,k\,l)$ plane of a cubic lattice of unit cell length a, can be shown to be: $d_{hkl} = \dfrac{a}{\sqrt{h^2 + k^2 + l^2}}$.

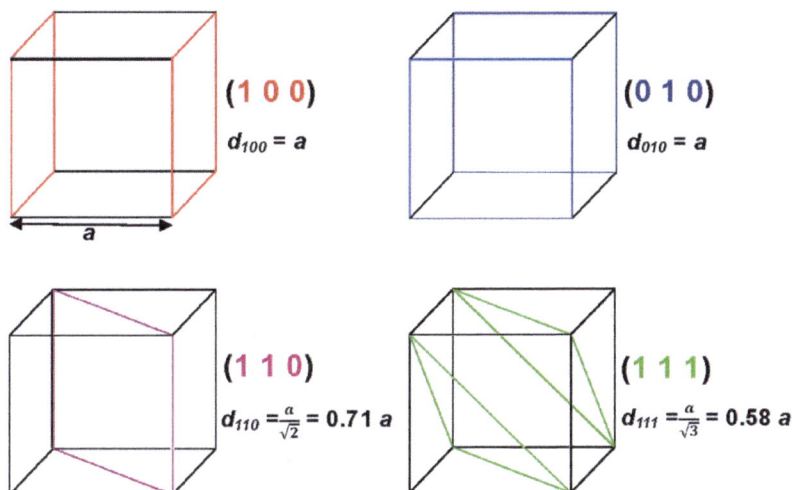

Figure 1.14 Miller planes in a 3-D cubic lattice, with unit cell of length *a*. Distances between the planes are shown.

1.4 X-ray Diffraction

X-ray diffraction is the most popular and efficient technique to determine precisely, the structure of crystals. X-rays are electromagnetic waves with wavelength typically in the range of ~0.5–10 Ångströms ($1 \text{ Å} = 10^{-10}$ m). **Waves** can be represented mathematically as a function of displacement, x and time, t:

$$f(x, t) = A \sin \left[2\pi \left(\frac{x}{\lambda} - vt \right) \right]$$

where A = amplitude, λ = wavelength, v = frequency. At any time t, $f(x)$ can be visualized as shown in Figure 1.15(a). Superposition of waves is shown schematically using the cases where they are in phase and out of phase by $\pi/2$ or π (Figure 1.15(b)–(d)).

1.4.1 Interference of Waves

Waves with different path lengths in a medium and hence relative phases (phase differences) interfere, leading to the enhancement or annihilation as shown in Figure 1.15(e); this follows directly from the illustration in Figure 1.15(b)–(d).

When shifted by integral multiples of λ, the waves add to each other fully, and when shifted by half-integer multiples of λ, they annihilate

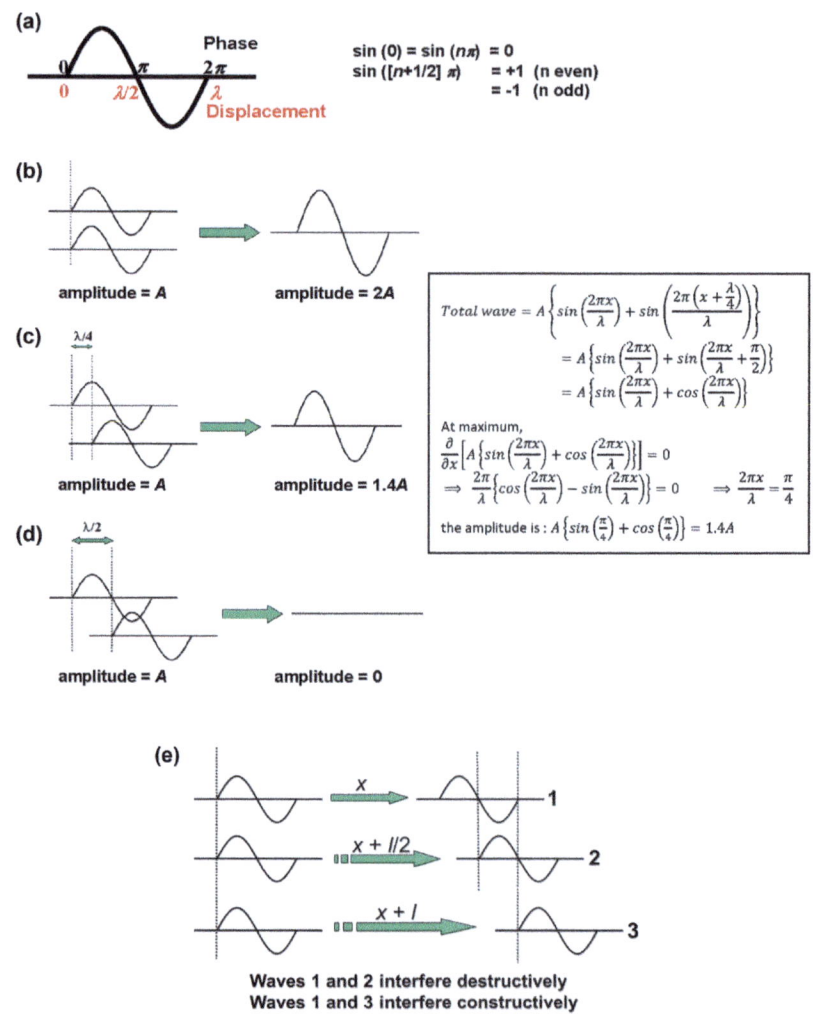

Figure 1.15 (a) Representation of a sinusoidal wave. Superposition of waves leading to (b) constructive, (c) partially destructive and (d) fully destructive interference; estimation of the total amplitude in (c) is shown in the inset. (e) Interference of waves traveling through different path lengths.

each other fully. Shifts of intermediate magnitude will lead to partial destructive interference.

1.4.2 Bragg's Law for Diffraction

Bragg's law is essentially the phenomenon described in Section 1.4.1. in action. X-ray (could equally well be electron or neutron) waves

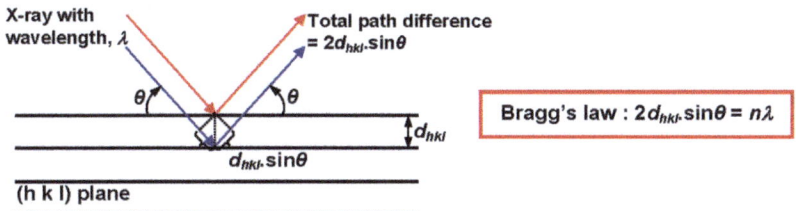

Figure 1.16   Derivation of Bragg's law for X-ray diffraction; n is an integer, and is called the order of reflection.

reflected from parallel planes ($h\ k\ l$) of a crystal undergo interference, with non-zero intensity emerging only when the path difference between the two is an integral multiple of the wavelength, λ (Figure 1.16); the condition for this is related to the angle of incidence, θ and the interplanar distance, d_{hkl}.

- It is an interesting question to consider what happens if the beam deviates slightly from the Bragg angle, θ; will there be partial interference or complete destructive interference? (This has relevance to line broadening in very small crystals).
- Another interesting point to note: the Bragg's law equation shows that a reflection for $n = 2$ is equivalent to an $n = 1$ reflection from a Miller plane with spacing of $d_{hkl}/2$; therefore, in practice one always works with just first order ($n = 1$) reflections.

1.4.3   Powder X-ray Diffraction

The powder X-ray diffraction experiment involves scattering a monochromatic X-ray beam from a collection of randomly oriented microcrystals. The diffracted beams satisfying the Bragg condition for each Miller plane will form a cone (due to the random orientation of the microcrystals) at scattering angle, 2θ with respect to the incident beam (Figure 1.17).

A typical example of a powder **X-ray diffraction pattern** is provided in Figure 1.18. The pattern corresponds to NaCl; incidentally, the first crystal structure determination by X-ray diffraction (Bragg and Bragg) was that of NaCl.

- The basic analysis of the powder X-ray diffraction pattern involves fitting the observed peak positions (in terms of the 2θ values) to Miller planes corresponding to a unit cell belonging to a specific crystal system. The process is called **indexing**.

(a)

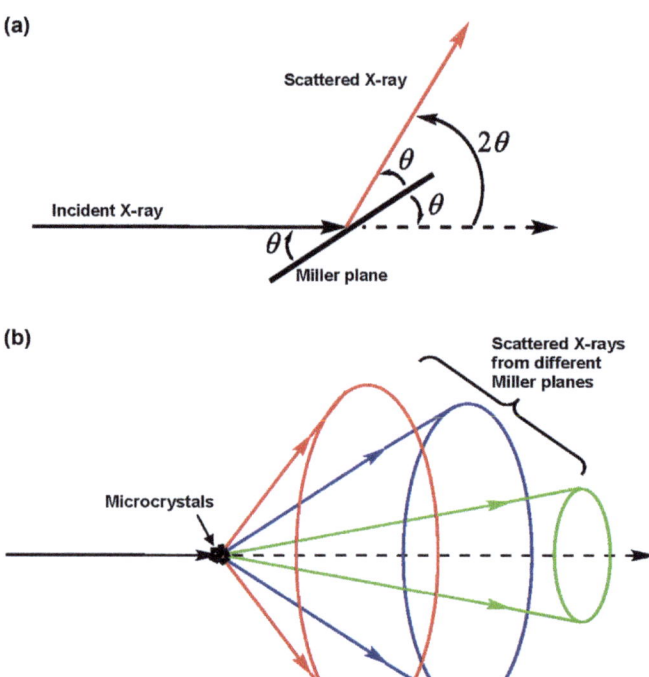

(b)

Figure 1.17   (a) Scattering of an X-ray beam from a Miller plane satisfying the Bragg condition at θ, causes deviation from its path by 2θ; (b) X-rays Bragg scattered from different Miller planes in a randomly oriented collection of microcrystals form cones.

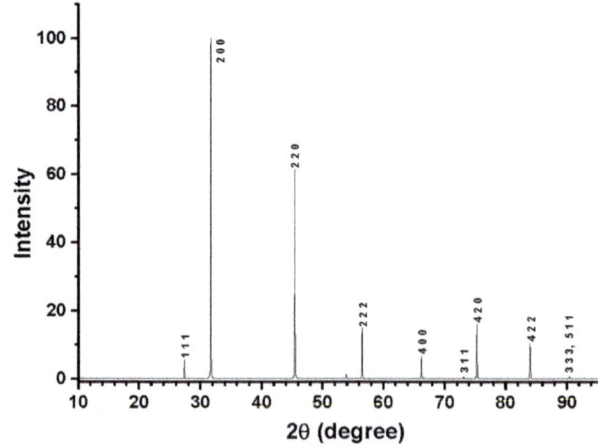

Figure 1.18   Powder X-ray diffraction pattern of NaCl.

- The different crystal systems, usually starting with the highest symmetry one (cubic), are explored. In each case, there is a relationship between the unit cell parameters, the Miller plane indices $(h\,k\,l)$ and the interplanar spacing d_{hkl}; the equation for the cubic system, $d_{hkl} = \dfrac{a}{\sqrt{h^2 + k^2 + l^2}}$ (Section 1.3.7) is the simplest.
- Indexing involves the determination of a common set of a, b, c, α, β, γ within a chosen system that satisfies the observed diffraction peaks (just the 2θ values), so that at the end of this process, one knows the crystal system and unit cell dimensions of the crystal under investigation. Careful analysis of the indexing can also reveal the relevant Bravais lattice, as will be demonstrated in the following sections.

The essential idea of indexing can be described using the NaCl diffraction data, as shown in Table 1.5.

- Using the 2θ values from the experimental diffraction pattern and the X-ray wavelength, the corresponding d values are calculated (n in the Bragg equation is taken as 1; see Section 1.4.2).
- Various combinations of integral values of h, k, and l (starting with, say, $(1\,0\,0)$) are tried for each d value to get the same value of $a = d\sqrt{(h^2 + k^2 + l^2)}$ (Section 1.3.7) within acceptable numerical deviations (this is essentially a statistical fitting protocol).
- In the NaCl case, this works out fine as the crystal does belong to a cubic system.
- If the crystal does not belong to the cubic system, then this fitting would never be satisfactory, and one would have to move on to

Table 1.5 Indexing of the NaCl powder X-ray diffraction data; the X-ray source is CuK$_\alpha$ ($\lambda = 1.5406$ Å).

2θ(deg.)	d(Å)	h	k	l	$(h^2 + k^2 + l^2)^{\frac{1}{2}}$	a(Å)
27.367	3.256	1	1	1	1.732	5.639
31.704	2.820	2	0	0	2.000	5.640
45.448	1.994	2	2	0	2.828	5.639
53.869	1.700	3	1	1	3.317	5.639
56.473	1.628	2	2	2	3.464	5.639
66.227	1.410	4	0	0	4.000	5.640
73.071	1.294	3	3	1	4.359	5.641
75.293	1.261	4	2	0	4.472	5.639
83.992	1.151	4	2	2	4.899	5.639
90.416	1.085	3	3	3	5.196	5.638
90.416	1.085	5	1	1	5.196	5.638

the next system, say tetragonal, and try a similar fitting, now with two unit cell parameters a and c.

- The process can go on until the triclinic system where the six independent unit cell parameters will need to be fitted.
- A careful observation reveals that in Table 1.5, all the ($h\ k\ l$) sets contain either all odd or all even numbers; cross combinations (for example, (1 0 0)) are absent. This is not accidental, and is determined by the Bravais lattice as explained below.

1.4.4 Systematic Absence

The X-ray diffraction patterns observed for crystals with primitive cubic, body-centered cubic and face-centered cubic lattices (Figure 1.19) demonstrate the idea of systematic absences.

- The primitive cube with only the vertices of the cube as lattice points, shows the diffraction from all the possible Miller planes.
- The systematic absences in the bcc and fcc lattices can be visualized as arising due to the destructive interference between

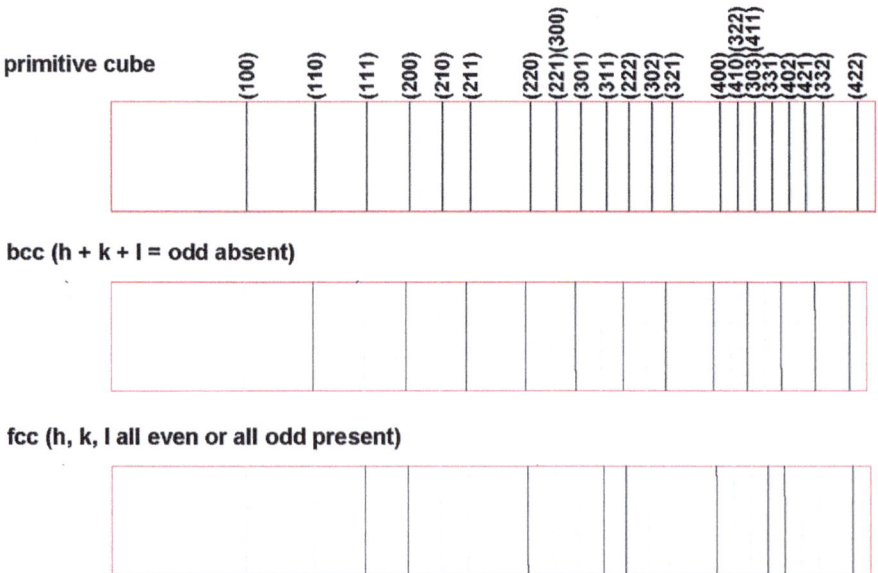

Figure 1.19 Schematic X-ray diffraction pattern of crystals with primitive cube, bcc, and fcc lattices; systematic absences in the latter two are indicated by the light blue lines. The systematic absence conditions are stated.

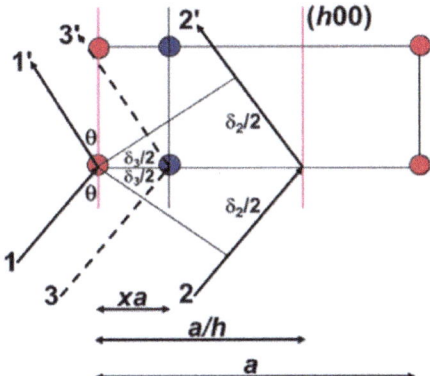

Figure 1.20 Schematic diagram showing X-ray scattering from two atoms (red and blue) of a basis, satisfying the Bragg condition for reflection from (*h* 0 0) plane.

the X-rays scattered from the lattice points at the vertices and at the body center/face centers respectively.

- A deeper look at the interference of waves scattered from specifically related lattice positions is required to understand this phenomenon better.
- The idea of **structure factor** discussed in the following section provides a general framework to understand this.

1.4.5 Structure Factor

A physical approach to construct the structure factor for the X-ray scattering from a Miller plane (*h k l*) involves the path and hence phase difference between the X-rays (satisfying the Bragg condition for that plane) reflected by the different atoms of the basis. The atom positions are well-defined with respect to each other. A different approach will be discussed later in Section 1.4.9. As an illustrative case, we first look at the 1-D problem with a unit cell of length *a* (Figure 1.20) along the *x*-axis.

- $d_{100} = a$; for plane (*h* 0 0) that is parallel to (1 0 0), $d_{h00} = \dfrac{a}{h}$.
- The reflection from (*h* 0 0) is equivalent to the *h*th order reflection from (1 0 0) (Section 1.4.2); this follows from Bragg's law:

$$2d \cdot \sin\theta = n\lambda \Rightarrow 2\frac{d}{n} \cdot \sin\theta = \lambda.$$

- Path difference between the waves 1 and 2 (1′ 2′) satisfying the Bragg condition, $\Delta l = \delta_2 = \lambda$.

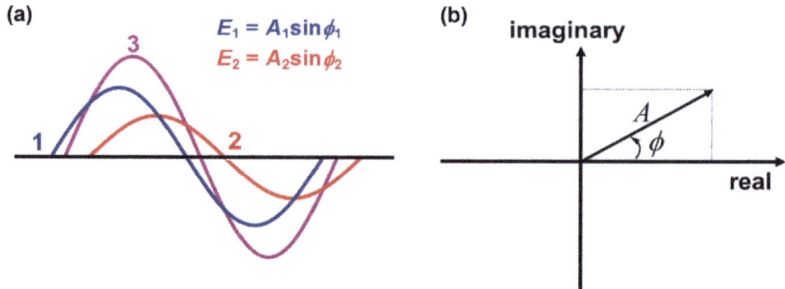

Figure 1.21   (a) Addition of two waves with the same frequency, but different amplitude and phase. (b) Representation of a wave as a vector in complex space.

- Using similar triangles, the path difference between the waves 1 and 3 (1′ 3′), $\Delta l = \delta_3 = \delta_2 \left(\dfrac{xa}{a/h} \right) = \lambda hx$.
- Phase difference for 1′ 3′, $\Delta \varphi = \left(\dfrac{2\pi}{\lambda} \right) \lambda hx = 2\pi hx$ [note: $\Delta l = \lambda \Rightarrow \Delta \varphi = 2\pi$].

For the 3-D case, phase difference between the two diffracted X-rays can be worked out in a similar way to give $\Delta \varphi = 2\pi\,(hx + ky + lz)$. Two waves with the same frequency, scattered from different atomic layers with phases φ_1 and φ_2, and amplitudes A_1 and A_2 (for example, if the atoms are not the same) can add up to form a wave 3 as shown in Figure 1.21(a).

- Scattered X-ray intensity will be due to the sum of the contributions from the different individual waves.
- A wave can be represented as a vector in complex space (Figure 1.21(b)), written as $A(\cos\varphi + i\sin\varphi) = Ae^{i\varphi}$

A measure of the scattering amplitude of the X-ray from an atom is the **atomic scattering factor**, defined as:

$$f = \frac{\text{amplitude of wave scattered by an atom}}{\text{amplitude of wave scattered by one electron}}.$$

This is clearly related to the electron density on the atom. X-ray wave scattered from an atom, n (scattering factor, f_n), located at a position with coordinates x_n, y_n, z_n, will contribute to the structure factor for the Miller plane ($h\ k\ l$) with phase $2\pi(hx_n + ky_n + lz_n)$ and amplitude f_n. The **structure factor** can now be written as:

$$S_{hkl} = \sum_n f_n e^{2\pi i(hx_n + ky_n + lz_n)}$$

where n represents the atoms in the unit cell. In this expression, information on the atom type comes from f_n and the atom position, from the coordinates x_n, y_n, z_n. The physical significance of the structure factor is that the intensity of the scattered X-ray from the $(h\,k\,l)$ plane is:

$I_{hkl} \propto S^*_{hkl} \cdot S_{hkl}$ (its utilization will be discussed further in Section 1.4.10)

The relevance of the structure factor can be illustrated by using it to understand the systematic absence in a bcc lattice as follows:

- Imagine the bcc lattice unit cell to be a simple (primitive) cubic cell with two atoms in the basis, A at position $(0\,0\,0)$ and B at $\left(\dfrac{1}{2}\dfrac{1}{2}\dfrac{1}{2}\right)$; these are the fractional coordinates expressing the position with respect to the unit cell axis.
- $S_{hkl} = f_A + f_B e^{2\pi i\left(\frac{h}{2} + \frac{k}{2} + \frac{l}{2}\right)} = f_A + f_B e^{\pi i(h+k+l)}$.
- As atoms at A and B are the same in a bcc lattice, $f_A = f_A = f$, and therefore,
  $$S_{hkl} = f\left(1 + e^{\pi i\,(h+k+l)}\right)$$
- As $e^{n\pi i} = \pm 1$ for n even/odd, the equation for the structure factor shows that
  $S_{hkl} = 2f$ when $h+k+l$ is even
  $S_{hkl} = 0$ when $h+k+l$ is odd
- This explains the absence of X-ray scattering from those Miller planes which satisfy $h+k+l = $ odd in a bcc lattice.

1.4.6   von Laue Condition for X-ray Diffraction

The Bragg's law for X-ray diffraction (Section 1.4.2) can be recast in the language of wave vectors and translations in momentum space (reciprocal lattice space) obtained by the Fourier transform of the periodic real lattice space. The meaning of the latter will become clearer through the following discussion.

A wave can be described by a wave vector, defined as $\vec{k} = \dfrac{2\pi}{\lambda}\hat{n}$, where λ is the wavelength and $\hat{n}$ is the unit vector along the direction of propagation of the wave. Note that $\vec{k}$ represents a momentum vector, since $\hbar\vec{k} = \dfrac{h}{\lambda}\hat{n} = \vec{p}$. Consider the X-ray scattering in a manner very similar to that described in Figure 1.16, but now represented using the incident wave vector, $\vec{k}$ and scattered wave vector, $\vec{k'}$ (from a pair of atoms at positions related by a lattice vector, $\vec{d}$); this is illustrated in Figure 1.22.

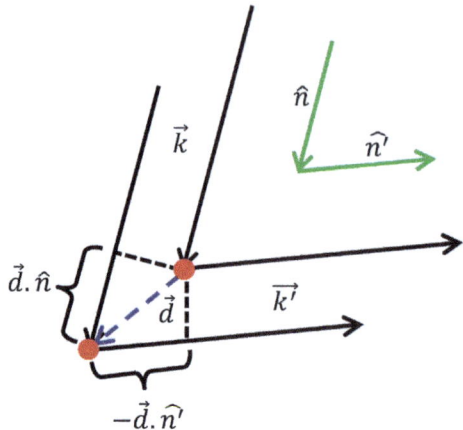

Figure 1.22 Schematic diagram of X-ray scattering from the lattice atoms (red filled circles) used to derive the von Laue condition for constructive interference.

- Projection of the vector $\vec{d}$ on the vector $\vec{k}$ (direction of which is represented by the unit vector $\hat{n}$) is given by $\vec{d} \cdot \hat{n}$; similarly $\vec{d}$ on the vector $\vec{k}'$ is given by $-\vec{d} \cdot \hat{n}'$.
- The total path difference of the second wave with respect to the first one is given by $\Delta l = \vec{d} \cdot \hat{n} - \vec{d} \cdot \hat{n}' = \vec{d} \cdot (\hat{n} - \hat{n}')$.
- Since $\vec{k} = \dfrac{2\pi}{\lambda} \hat{n}$ and $\vec{k}' = \dfrac{2\pi}{\lambda} \hat{n}'$, $\Delta l = \dfrac{\lambda}{2\pi} \vec{d} \cdot \left(\vec{k} - \vec{k}'\right) = \dfrac{\lambda}{2\pi} \vec{d} \cdot \Delta \vec{k}$.
- For constructive interference, $\Delta l = m\lambda$ where m is an integer.
- Therefore, $m\lambda = \dfrac{\lambda}{2\pi} \vec{d} \cdot \left(\vec{k} - \vec{k}'\right) \Rightarrow \vec{d} \cdot \Delta \vec{k} = 2\pi m$.
- If $\vec{R}$ represents a general translation vector in the Bravais lattice, the condition for constructive interference can be generalized as: $\vec{R} \cdot \Delta \vec{k} = 2\pi m$.
- Equivalently: $e^{i\vec{R} \cdot \Delta \vec{k}} = 1$, as $e^{2\pi m i} = [\cos(2m\pi) + i \sin(2m\pi)] = 1$.

Introducing the **reciprocal lattice**: Similar to the Bravais lattice vector for a 3-D lattice, $\vec{R} = n_1 \vec{a}_1 + n_2 \vec{a}_2 + n_3 \vec{a}_3$ (Section 1.3.5), a reciprocal lattice vector, $\vec{K} = h\vec{b}_1 + k\vec{b}_2 + l\vec{b}_3$ can be defined. The primitive vectors in reciprocal space are a set of orthonormal vectors for the primitive (real) lattice vectors, $\vec{a}_1, \vec{a}_2, \vec{a}_3$, defined as follows:

$$\vec{b}_1 = 2\pi \frac{\vec{a}_2 \times \vec{a}_3}{\vec{a}_1 \cdot \vec{a}_2 \times \vec{a}_3} \qquad \vec{b}_2 = 2\pi \frac{\vec{a}_3 \times \vec{a}_1}{\vec{a}_1 \cdot \vec{a}_2 \times \vec{a}_3} \qquad \vec{b}_3 = 2\pi \frac{\vec{a}_1 \times \vec{a}_2}{\vec{a}_1 \cdot \vec{a}_2 \times \vec{a}_3}$$

- This definition ensures that $\vec{R} \cdot \vec{K} = 2\pi m$, where m is an integer (this is easily proved; try).

- Comparing this to the previous equation, $\vec{R} \cdot \Delta \vec{k} = 2\pi m$, we get the final von Laue condition for constructive interference:

$$\Delta \vec{k} = \vec{K}.$$

- The physical meaning of this relation is that constructive interference occurs when the change in the X-ray wave vector is equal to one of the reciprocal lattice vectors.
- Mathematically, the reciprocal lattice is the Fourier transform of the Bravais lattice. The latter is referred to as the real lattice as it represents the real spatial arrangement of the atoms in a crystal; reciprocal lattice can be visualized as the lattice in momentum space, and the Laue condition as the conservation of crystal momentum.

The meaning and construction of reciprocal lattice are discussed briefly in the next section, before returning to the X-ray diffraction problem.

1.4.7 Reciprocal Lattice

As noted above, the reciprocal lattice vectors form a set of orthonormal vectors for the real lattice vectors. This can be shown geometrically using first, a 2-D Bravais lattice.

- In the oblique lattice shown in Figure 1.23(a), $\vec{a}_1$ and $\vec{a}_2$ are the real lattice vectors; the unit cell defined by these vectors is shown as the green parallelogram.
- The perpendiculars to $\vec{a}_2$ and $\vec{a}_1$ give the axial directions of $\vec{b}_1$ and $\vec{b}_2$ respectively; this ensures the orthogonality criteria, $\vec{b}_1 \perp \vec{a}_2$ and $\vec{b}_2 \perp \vec{a}_1$.

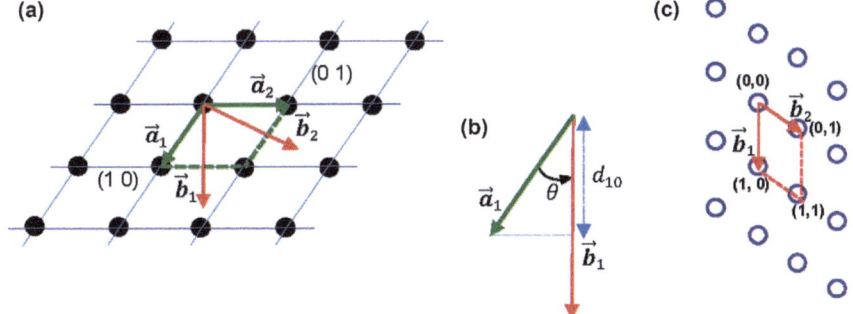

Figure 1.23 (a) A 2-D oblique lattice showing the primitive lattice vectors, $\vec{a}_1$ and $\vec{a}_2$, the Miller planes (1 0) and (0 1) and the direction of the reciprocal lattice vectors $\vec{b}_1$ and $\vec{b}_2$. (b) Geometry to calculate $\vec{a}_1 \cdot \vec{b}_1$. (c) Reciprocal lattice of the oblique lattice defined by the reciprocal lattice vectors $\vec{b}_1$ and $\vec{b}_2$.

- Clearly, $\vec{b}_1$ is perpendicular to the (1 0) plane (this is to be imagined, as the reciprocal vector is in inverse space whereas the Miller planes are in real space), and $\vec{b}_2$ is perpendicular to the (0 1) plane. This shows that each reciprocal lattice vector is associated uniquely to a Miller plane.
- The magnitude of $\vec{b}_1$ is chosen to be $\dfrac{2\pi}{d_{10}}$ and of $\vec{b}_2$ to be $\dfrac{2\pi}{d_{01}}$; this ensures that $\vec{a}_1 \cdot \vec{b}_1 = |\vec{a}_1||\vec{b}_1|\cos\theta = (|\vec{a}_1|\cos\theta)|\vec{b}_1| = d_{10} \times$ $\dfrac{2\pi}{d_{10}} = 2\pi$ (Figure 1.23(b) shows that $|\vec{a}_1|\cos\theta = d_{10}$, the spacing for the (1 0) planes). Similarly, $\vec{a}_2 \cdot \vec{b}_2 = 2\pi$.
- Based on the above definitions, it is clear that the reciprocal lattice vector corresponding to a Miller plane with larger spacing will be shorter and *vice versa*.
- Using the reciprocal lattice vectors, $\vec{b}_1$ and $\vec{b}_2$ we construct the reciprocal lattice (blue) shown in Figure 1.23(c); each reciprocal lattice point can be labeled by the coordinates (h, k) corresponding to the Miller plane $(h\ k)$ associated with it. The point $(1, 1)$ in the reciprocal lattice shows that the vector $(\vec{b}_1 + \vec{b}_2)$ is perpendicular to the (1 1) plane in the real lattice.

Similarly one can construct geometrically, the reciprocal lattice for 3-D lattices, as follows:

- Figure 1.24 shows a triclinic lattice cell with primitive vectors $\vec{a}_1, \vec{a}_2, \vec{a}_3$.
- From the definition of the reciprocal vector $\vec{b}_3$, it is clear that it is perpendicular to the plane defined by $\vec{a}_1$ and $\vec{a}_2$; the magnitude of this vector (in reciprocal length, of course) is given by:

$$|\vec{b}_3| = 2\pi \left|\frac{\vec{a}_1 \times \vec{a}_2}{\vec{a}_1 \cdot \vec{a}_2 \times \vec{a}_3}\right| = 2\pi \frac{\text{area of base}}{\text{unit cell volume}} = \frac{2\pi}{\text{height}} = \frac{2\pi}{d_{001}}.$$

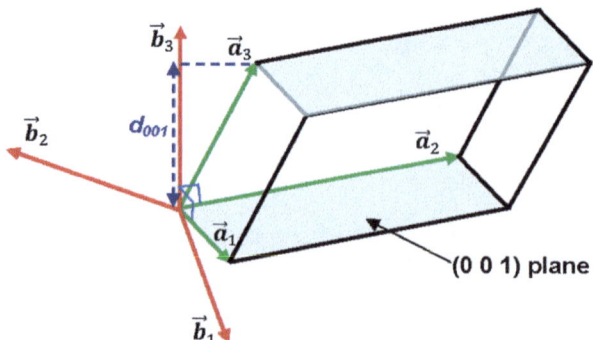

Figure 1.24 Constructing geometrically, the reciprocal lattice vectors of a 3-D triclinic Bravais lattice.

- Similarly the reciprocal vectors, $\vec{b}_1$ and $\vec{b}_2$ can be visualized.

The reciprocal lattice vectors can be derived analytically. Using the definition given in Section 1.4.6 (and shown geometrically in this section), the primitive reciprocal lattice vectors of some typical 3-D Bravais lattices (cubic lattices) can be derived as follows.

- Primitive cube: $\vec{a}_1 = a\hat{x}$; $\vec{a}_2 = a\hat{y}$; $\vec{a}_3 = a\hat{z}$.
- $\vec{b}_1 = 2\pi \dfrac{a\hat{y} \times a\hat{z}}{a\hat{x} \cdot a\hat{y} \times a\hat{z}} = 2\pi \dfrac{a^2\hat{x}}{a^3} = \dfrac{2\pi}{a}\hat{x}.$
- Similarly, $\vec{b}_2 = \dfrac{2\pi}{a}\hat{y}$ and $\vec{b}_3 = \dfrac{2\pi}{a}\hat{z}.$
- Clearly, the primitive reciprocal lattice vectors, $\vec{b}_1$, $\vec{b}_2$, $\vec{b}_3$ of a simple cubic Bravais lattice form a lattice with cubic symmetry in inverse space.
- For a simple cubic lattice of unit cell length, a Å, the reciprocal lattice is cubic with unit cells of length $\dfrac{2\pi}{a}$ Å^{-1}.

Reciprocal lattice of the fcc lattice can be shown to be a lattice with bcc symmetry!

- $\vec{a}_1 = \dfrac{a}{2}(\hat{y} + \hat{z})$; $\vec{a}_2 = \dfrac{a}{2}(\hat{z} + \hat{x})$; $\vec{a}_3 = \dfrac{a}{2}(\hat{x} + \hat{y})$

- $\vec{b}_1 = 2\pi \dfrac{\left(\dfrac{a}{2}\right)^2}{\left(\dfrac{a}{2}\right)^3} \dfrac{(\hat{z} + \hat{x}) \times (\hat{x} + \hat{y})}{(\hat{y} + \hat{z}) \cdot (\hat{z} + \hat{x}) \times (\hat{x} + \hat{y})}$

$\quad = \dfrac{4\pi}{a} \dfrac{(\hat{y} + \hat{z} - \hat{x})}{(\hat{y} + \hat{z}) \cdot (\hat{y} + \hat{z} - \hat{x})} = \dfrac{2\pi}{a}(\hat{y} + \hat{z} - \hat{x})$

- Similarly, $\vec{b}_2 = \dfrac{2\pi}{a}(\hat{z} + \hat{x} - \hat{y})$, and $\vec{b}_3 = \dfrac{2\pi}{a}(\hat{x} + \hat{y} - \hat{z})$

The reverse is also true; the reciprocal lattice of the bcc lattice has fcc symmetry; this is illustrated in Figure 1.25. Note that the

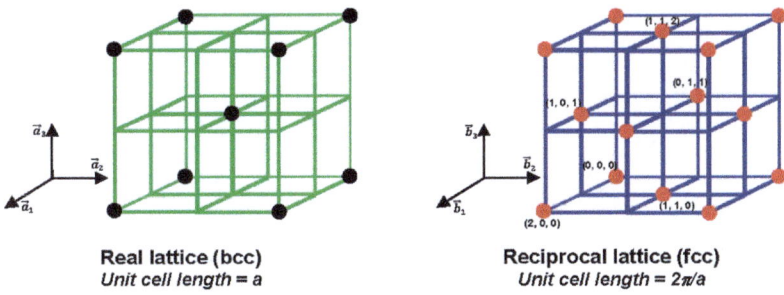

Real lattice (bcc)
Unit cell length = a

Reciprocal lattice (fcc)
Unit cell length = 2π/a

Figure 1.25 Real bcc lattice and the corresponding reciprocal fcc lattice.

reciprocal lattice points correspond to the allowed Miller indices of the bcc lattice, *i.e.*, $h+k+l=$ odd are absent.

- $\vec{a}_1 = \dfrac{a}{2}(\hat{y}+\hat{z}-\hat{x}); \; \vec{a}_2 = \dfrac{a}{2}(\hat{z}+\hat{x}-\hat{y}); \; \vec{a}_3 = \dfrac{a}{2}(\hat{x}+\hat{y}-\hat{z})$

- $\vec{b}_1 = \dfrac{2\pi}{a}(\hat{y}+\hat{z}); \; \vec{b}_2 = \dfrac{2\pi}{a}(\hat{z}+\hat{x}); \; \vec{b}_3 = \dfrac{2\pi}{a}(\hat{x}+\hat{y})$

1.4.8   Ewald Construction

Recall the condition for constructive interference of scattered X-ray waves, $\Delta\vec{k}=\vec{K}$ (Section 1.4.6). This leads directly to the geometric construction illustrating the relation between the incident and scattered X-ray waves, the relevant diffraction angle, and the reciprocal lattice.

- $\vec{K}=h\vec{b}_1+k\vec{b}_2+l\vec{b}_3$ is any reciprocal lattice vector; it is uniquely associated with the (h k l) Miller plane in the real lattice, and $|\vec{K}|=\dfrac{2\pi}{d_{hkl}}$ (Section 1.4.7).
- The von Laue condition can be visualized geometrically using the **Ewald construction** in reciprocal space (Figure 1.26).
- Placing the incident X-ray wave vector, $\vec{k}$ on a lattice point in reciprocal space (chosen as (0, 0, 0) in the figure), and using it as the radius, a circle (sphere in 3-D) is drawn. If the circle passes

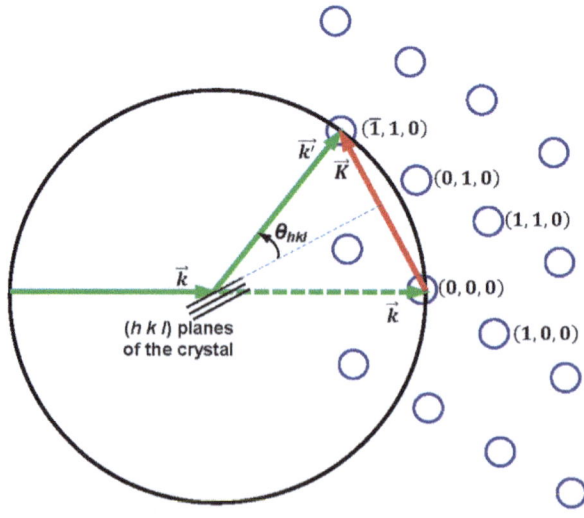

Figure 1.26   Ewald construction in reciprocal lattice (blue circles) showing the von Laue condition for constructive interference of X-rays; scattering from the real lattice plane is combined to show how Bragg's law can be inferred from the same picture.

through any other reciprocal lattice point, then it satisfies the condition, $\vec{k} + \vec{K} = \vec{k'}$, i.e., $\vec{k'} - \vec{k} = \vec{K}$ (note that the scattering is elastic; the wavelength and hence the magnitude of the wave vector are the same).

- Combining the diffraction picture in the real lattice (Figure 1.26), one can see that this is consistent with Bragg's law

$$\sin\theta_{hkl} = \dfrac{\frac{1}{2}|\vec{K}|}{|\vec{k'}|} = \dfrac{\frac{1}{2}\left(\frac{2\pi}{d_{hkl}}\right)}{\frac{2\pi}{\lambda}} = \dfrac{\lambda}{2d_{hkl}} \Rightarrow 2d_{hkl}\sin\theta_{hkl} = \lambda$$

$$\Rightarrow 2d\sin\theta_{hkl} = n\lambda$$

as every observed d can be related to an integral multiple of a d_{hkl}.

1.4.9 Structure Factor in Terms of the Reciprocal Lattice Vector

The structure factor can be derived based on the fact that there is interference between the X-rays scattered by the different atoms in the basis (Section 1.4.5). In the derivation of the von Laue condition (Section 1.4.6) it was seen that the path length difference is given by $\dfrac{\lambda}{2\pi}\vec{d} \cdot \Delta\vec{k} = \dfrac{\lambda}{2\pi}\vec{d} \cdot \vec{K}$. This translates to a phase difference of $\vec{d} \cdot \vec{K}$ (if the path difference is λ, the phase difference is 2π). The contribution of the phase factor to the wave can be written as $e^{i\vec{d} \cdot \vec{K}}$.

Summing up the contributions from all the atoms in the basis, the **structure factor** can be written as: $S_K = \sum_n f_n e^{i\vec{d}_n \cdot \vec{K}}$, where f_n and d_n are the atomic scattering factor and position vector of atom n in the unit cell respectively.

The above equation is applied below, to determine the structure factor and the systematic absence in the bcc lattice.

- As in Section 1.4.5, consider the bcc lattice unit cell to be a simple cubic cell with two atoms in the basis, 1 at position (0 0 0) and 2 at $\left(\dfrac{1}{2}\dfrac{1}{2}\dfrac{1}{2}\right)$. Therefore, $\vec{d}_1 = 0$ and $\vec{d}_2 = \dfrac{a}{2}(\hat{x} + \hat{y} + \hat{z})$; a is the length of the unit cell.
- The reciprocal lattice vector corresponding to an $(h\ k\ l)$ plane of the cubic lattice considered above, is $\vec{K} = \dfrac{2\pi}{a}(h\hat{x} + k\hat{y} + l\hat{z})$
- If the two atoms have the same scattering factor, f, $S_K = S_{hkl} = f\left[1 + e^{i\frac{a}{2}(\hat{x}+\hat{y}+\hat{z})\cdot\frac{2\pi}{a}(h\hat{x}+k\hat{y}+l\hat{z})}\right] = f\left[1 + e^{\pi i(h+k+l)}\right]$; this is the same

relation as derived in Section 1.4.5 and the systematic absence conditions follow as discussed there.

1.4.10 Basic Concepts of X-ray Structure Solution and Refinement

The essential concepts and steps involved in the actual X-ray diffraction data analysis and determination of the crystal structure can be summarized as follows:

- The intensity of the X-rays scattered from an $(h\ k\ l)$ plane, $I_{hkl} = S^*_{hkl}\ S_{hkl}$.
- From the intensities measured from different planes in a single crystal X-ray diffraction experiment (corrected for experimental factors such as absorption, polarization, reduction in intensity due to lattice vibration and incoherent scattering, *etc.*), the structure factors are determined. This involves the tricky issue of deciding the phase factor, the famous 'phase problem', discussed in specialized text books on crystallography (only $|S_{hkl}|$ is measured, but since $S_{hkl} = |S_{hkl}|e^{i\theta_{hkl}}$, the phase, θ_{hkl} is also required to determine the atom positions).
- As the atomic scattering factor is related to the electron density, one can use the picture of the spatial electron density distribution in the crystal, and replace the summation in the equation for the structure factor by the integral so that,
  $$S_K = \int f(\vec{r}) \cdot e^{i\vec{K}\cdot\vec{r}} d\vec{r}.$$
- Fourier transform leads to an electron density map called the Fourier map:
  $$\rho(\vec{r}) \propto f(\vec{r}) = \int S_K \cdot e^{-i\vec{K}\cdot\vec{r}} d\vec{K}.$$
- The Fourier map provides the initial structure solution; the total electron densities at specific points determine the type of atom present there.
- Using the initial solution, structure factors are calculated for each $(h\ k\ l)$ plane; this gives the calculated structure factor list, $S_{hkl}(\text{calc})$. From the experiment, one already has the experimental structure factor list, $S_{hkl}(\text{expt})$.
- The least squares method is used to carry out regression of $S_{hkl}(\text{calc})$ against $S_{hkl}(\text{expt})$. Quality of refinement is represented by the so-called 'R factor'.
- The model is revised iteratively to obtain decreasing R factor values. The final model used for the best $S_{hkl}(\text{calc})$ that gives the lowest 'R factor' is the refined structure.

1.5 Neutron and Electron Diffraction

Thanks to the wave–particle duality of neutrons and electrons, they can also be used to carry out diffraction experiments to analyze the structure of crystals. The basic principles governing the diffraction process are the same as those discussed for X-ray diffraction. From the de Broglie equation $p = \dfrac{h}{\lambda}$ (p = momentum, λ = wavelength, h = Planck constant), it is clear that the wavelengths of neutron or electron waves are determined by their momentum, and hence the velocity to which they are accelerated. Scattering of neutron or electron waves with wavelength of the order of Ångströms is used to carry out the diffraction study of single crystals; in high energy electron diffraction experiments much smaller wavelengths are extensively used.

1.5.1 Neutron Diffraction

A fundamental difference between X-ray and neutron scattering is that the former is determined by the electron density distribution around the atom nuclei, whereas the latter is directly by the nuclei and the magnetic moment at the atomic site. Because of the weak scattering from light atoms, determination of their positions in the crystal using X-rays is difficult.

- Neutron diffraction allows precise determination of the position of light atoms including H, even in the presence of heavier atoms.
- Isotopes can be distinguished in a neutron diffraction experiment.
- Since neutrons carry a magnetic moment, an ion present in a crystal with different magnetic moments at different sites, would scatter them differently; this allows the differentiation of such ions in the lattice. Neutron diffraction can therefore be used to reveal the magnetic structure of materials (see Section 5.4.5).

1.5.2 Electron Diffraction

Electron diffraction is often carried out together with electron microscopy. For example, a transmission electron microscope can be used to image nanomaterials and nanostructures, and also record electron diffraction from selected areas or features of the sample.

- A 100 kV electron beam would have a wavelength of ∼0.04 Å. The accelerating voltage is applied across the electron source (thermal or field emission) and an anode; the schematic diagram in

Figure 1.27 shows the electron beam path in a transmission electron microscope.

- The diffraction follows Bragg's law and a Fourier transform of the electron diffraction pattern provides a map of the corresponding Miller planes of the lattice structure. Since the electron beam wavelengths are often much smaller than those of typical X-rays used in diffraction studies, the Ewald sphere (Section 1.4.8) in electron diffraction experiments are much larger, sampling more reflections. As the curvature of the Ewald sphere is very small,

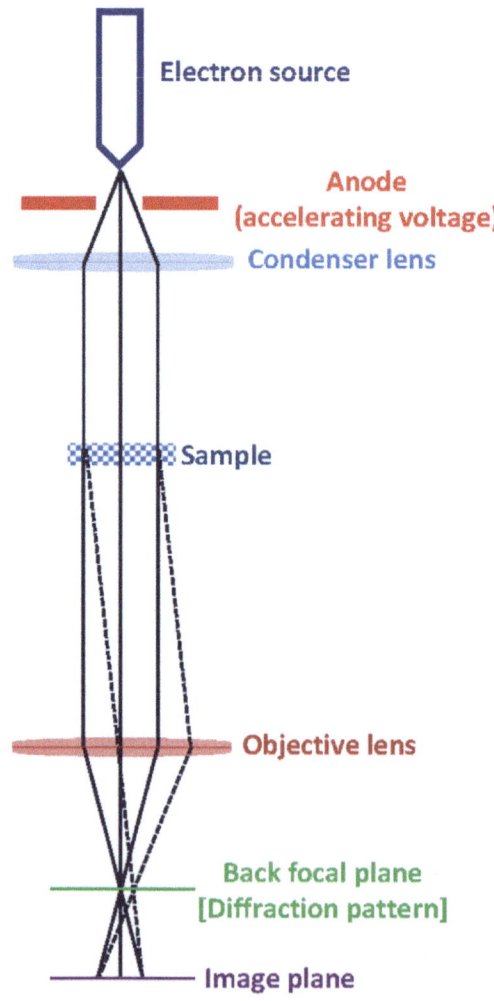

Electron source

Anode (accelerating voltage)

Condenser lens

Sample

Objective lens

Back focal plane [Diffraction pattern]

Image plane

Figure 1.27 Schematic diagram showing the electron source and the anode which accelerates the electrons, passage of the electron beam through the sample, and the formation of the image and diffraction pattern at the image and back focal planes respectively, in a transmission electron microscope; the broken lines indicate the scattered electron beams.

Miller indices corresponding to the reciprocal lattice points lying on the Ewald sphere can be read out directly using appropriate instrumentation.

- Using appropriate apertures, the diffraction pattern can be collected in the back focal plane as shown in Figure 1.27. The image is collected at the image plane as shown.
- The lenses are electromagnets which bend the electron beams; collection of electrons scattered from different positions produces the image, whereas electrons scattered at the same angle give the diffraction pattern.

1.6 Common Crystal Structure Motifs

It is important to be familiar with specific types of crystal structures which occur rather frequently or have unique and characteristic features. The structural motif has a direct bearing on the material's properties and functions. A few examples of general crystal structural classes selected at random are listed below with examples of specific materials that possess them.

- Rock salt : NaCl, FeO
- Fluorite : CaF_2, Li_2O
- Perovskite (ABO_3) : $BaTiO_3$
- Spinel (AB_2O_4) : $MgAl_2O_4$ (inverse spinel : Fe_3O_4)
- Rutile : TiO_2, NbO_2
- Diamond: C, Si
- Graphite : C
- Zinc blende : ZnS, GaAs
- Wurtzite : ZnS, AgI

It should be stressed that there are several more structural classes; it is worthwhile to consult textbooks on solid state chemistry, crystallography, *etc.*, for further examples.

1.7 Quasicrystals: A Brief Note

Quasicrystals are a fascinating phenomenon that has been unraveled over the last few decades. As shown in Section 1.3.1, translational periodicity ensures that rotational symmetry occurs only with orders 1, 2, 3, 4, and 6, leading eventually to the 230 space groups. Clearly, a crystal cannot have rotational symmetry of order 5, 7, 9, 10, *etc.* However, in 1982, Daniel Shechtman (Nobel Prize, 2011) found that

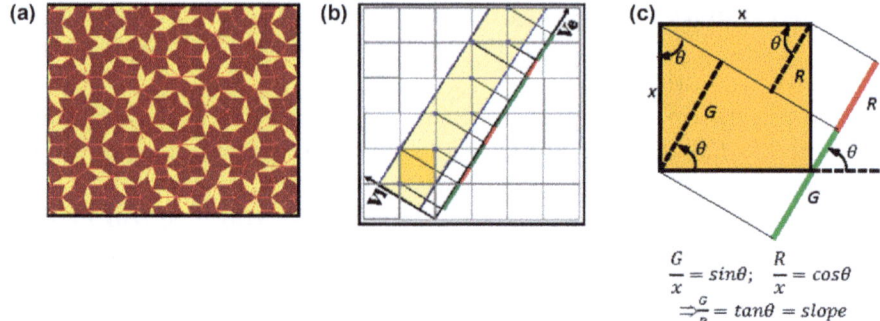

$$\frac{G}{x} = sin\theta; \quad \frac{R}{x} = cos\theta$$

$$\Rightarrow \frac{G}{R} = tan\theta = slope$$

Figure 1.28 (a) Penrose tiling; (b) a 1-D quasi-periodic sequence (Fibonacci chain) embedded in a higher dimensional (2-D) space in the form of a periodic square lattice. Adapted from ref. 1 with permission from Dr Steffen Weber, original author of the website. (c) Derivation of the G/R ratio.

electron diffraction of a rapidly cooled Al–Mn alloy produced a diffraction pattern that had 10-fold symmetry!

The term 'quasicrystal' refers to a structure that is ordered, but not periodic in the observable dimensions like 1-, 2-, or 3-D.

- The meaning of 'ordered' but not 'periodic' can be understood from figures like the famous Penrose tiling (Figure 1.28(a)). It is clearly an ordered structure with a 5-fold symmetry; obviously there cannot be, and there is no translational symmetry.
- The meaning of quasi-periodicity in a lower dimension mapped on to a periodic structure in a higher dimension can be understood from Figure 1.28(b).
  - The square lattice (periodic in 2-D) has a unit cell shaded deep yellow.
  - The points within a strip (light yellow) are projected on to an external space (V_e axis).
  - If the V_e axis has an irrational slope, the projection gives a quasi-periodic sequence (green (G) and red (R) segments in the ratio equal to the slope of the V_e axis (Figure 1.28(c)); for example, if the slope $= \dfrac{\sqrt{5}+1}{2} = 1.61803 \ldots$, the sequence is a Fibonacci chain (1-D quasi-periodic structure).
- Quasicrystals are such manifestations which are quasi-periodic in 2- or 3-D, but periodic in higher dimensions.

Reference

1. http://www.jcrystal.com/steffenweber/qc.html

2 Defects and Non-stoichiometry

2.1 Point, Line and Plane Defects

A general classification of defects in terms of their dimensionality is very useful. The terms, point, line, and plane defects indicate clearly the region over which a defect exists – a lattice or interstitial site, a whole row of lattice points or a plane formed by a 2-D array of lattice points. In a broad sense, the defect can be due to the absence of the atom/molecule at a specific site or the presence of an atom/molecule where it is not expected, based on the bulk structure.

- **Plane defect**: typical examples are small angle grain boundaries and stacking faults. The schematic diagrams in Figure 2.1 convey these concepts clearly.
- **Line defect**: the most common of these are edge and screw dislocations shown in Figure 2.2. The defect propagates parallel to the direction of the slip of the lattice planes in an edge dislocation; the defect propagation is perpendicular to the slip direction in a screw dislocation. The presence of a defect is shown by a mismatch (Burger's vector) if one follows an equal number of unit cell vectors in opposite directions to make a circuit around the defect area.

A wide range of **point defects** occur in crystals. They are broadly classified into intrinsic and extrinsic defects.

Core Concepts for a Course on Materials Chemistry
By T. P. Radhakrishnan
© T. P. Radhakrishnan 2023
Published by the Royal Society of Chemistry, www.rsc.org

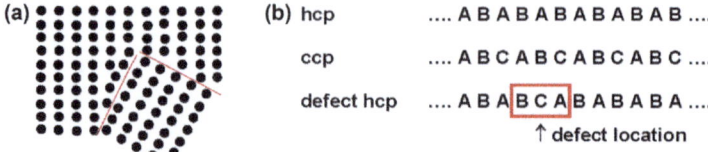

Figure 2.1   Plane defects: (a) small angle grain boundary (angle exaggerated for easier viewing), (b) stacking fault in a hexagonal close packing (hcp) sequence contrasted with defect-free hcp and ccp.

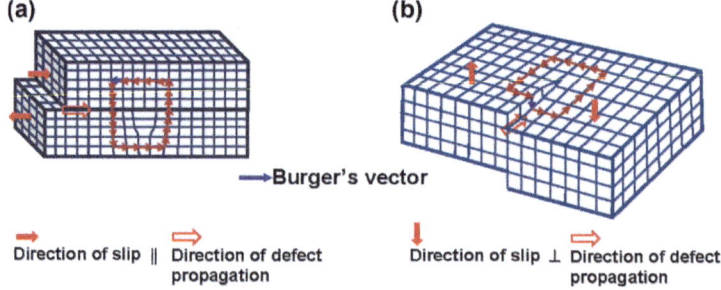

Figure 2.2   Schematic drawing of (a) edge and (b) screw dislocation.

- Intrinsic defects: vacancy, self-interstitial, Schottky, Frenkel.
- Extrinsic defects: substitutional/interstitial impurity, aliovalent impurity, charge compensation defect, non-stoichiometry, color centers.
- The defect formation is driven entirely by entropy factors as the enthalpy required is always positive.

  Intrinsic defects and some of the simple extrinsic ones are discussed in Sections 2.2 and 2.3. A more detailed discussion of non-stoichiometry and color centers follows in Sections 2.4 and 2.5.

2.2   Intrinsic Defects

Intrinsic defects arise when an atom or molecule is not present at the lattice site where it is expected based on the bulk crystal structure, or when it is present at a site or interstitial region (space between lattice points) where it is not expected to be.

There is usually an energy cost associated with the formation of the defect (missing or undesirable interactions); however, there is also an entropy gain due to the formation of the defect. The defect arises when an overall free energy (depending on the temperature of the system) gain is involved. The common types of intrinsic defects discussed below are:

- vacancy
- self-interstitial
- Schottky
- Frenkel

Defects exert a significant influence on several material properties such as electrical conductivity, optical characteristics, magnetism, and mechanical and thermal attributes. Some of these effects will be discussed in later sections that deal with the specific topics.

2.2.1 Vacancy

Vacancy is the simplest form of intrinsic defect, arising due to the absence of an atom or molecule at a lattice site where it is expected to be present; this is shown schematically in Figure 2.3(a). The number of vacancies, n formed with an average energy cost of $\bar{E}_v$ at temperature, T in a lattice with N sites, is given by:

$$n = Ne^{-\bar{E}_v/k_B T} \text{ where } k_B = \text{Boltzmann constant}$$

Temperature plays a crucial role in the formation of vacancies. The above equation shows that, for a typical crystal where the energy of formation of a vacancy is 1.0 eV mol^{-1}, the ratio of the number of vacancies at 1000 K to that at 500 K is $\dfrac{n_{1000}}{n_{500}} \sim 10^5$.

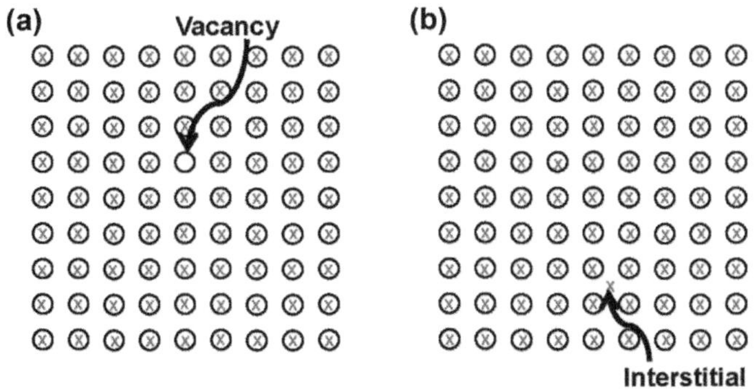

Figure 2.3 Point defects in the form of (a) vacancy and (b) self-interstitial; circles represent the lattice points and x the basis.

2.2.2 Self-interstitial

These defects form when an atom or molecule, rather than being at the lattice site, sits at an interstitial position as shown in Figure 2.3(b); this happens mostly in elemental solids.

2.2.3 Schottky Defect

Schottky defects form when cation–anion combinations are missing from their respective lattice sites (Figure 2.4(a)) leading to cation and anion vacancies maintaining charge balance. They occur often in alkali halide crystals. As the volume of the crystal remains unchanged but the mass is now reduced, these defects lead to a decrease in the density of the crystal. The defects can also be visualized as arising due to the migration of the cation–anion pair to the crystal surface. The defect population is again given by, $n = Ne^{-\bar{E}_s/2k_BT}$ ($\bar{E}_s$ is the average energy cost of the Schottky defect formation, and the factor of 2 arises due to the pair of ions involved). The density of a crystal can be estimated from the unit cell information. An example:

- consider a salt M^+X^- (molecular weight $= 74.6$ g mol^{-1}) with the rock salt structure (interpenetrating fcc lattices of M^+ and X^-) and the distance M^+-$X^- = 0.32$ nm
- the number of M^+X^- in the conventional unit cell $= 4$; unit cell length $= 0.64$ nm
- the density, $\rho = \dfrac{4 \times 74.6 \times 10^{-3}}{6.023 \times 10^{23} \times (0.64 \times 10^{-9})^3} = 1.890 \times 10^3$ kg m^{-3}

- if the above crystal has 0.1% Schottky defects, the density will be:

$$\rho_{\text{def}\cdot\text{cryst}} = 1.890 \times 10^3 \times \frac{99.9}{100} = 1.888 \times 10^3 \text{ kg m}^{-3}$$

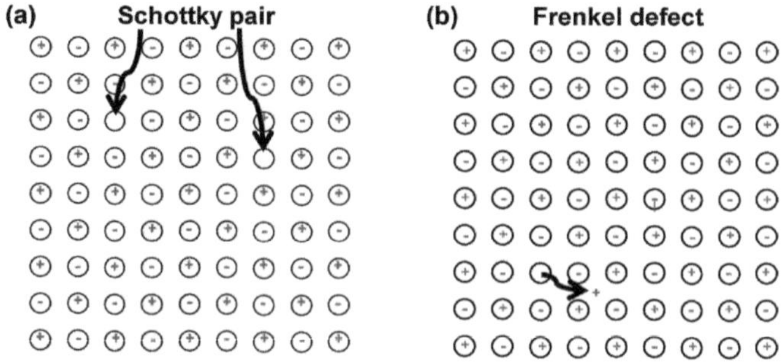

Figure 2.4 (a) Schottky and (b) Frenkel defects; circles represent the points in two interpenetrating lattices and + and − the ions that form the basis of each.

2.2.4 Frenkel Defect

Frenkel defects form when an ion (often the cation) migrates from the lattice site where it is expected to be, to an interstitial position (Figure 2.4(b)). These occur generally in relatively open structures with large interstitial spaces. Silver halides are typical crystals in which Frenkel defects are found. As the vacancy is compensated by the interstitial ion, Frenkel defects lead to no change in the density of the crystal. The defect population is given by, $n = \sqrt{(NN_I)}e^{-\bar{E}_{Frenkel}/2k_B T}$ (N_I is the number of interstitial sites).

2.3 Extrinsic Defects

Extrinsic defects are characterized by the presence of impurities or foreign atoms, ions, or molecules in a crystal lattice; the percentage of impurity atoms will always be well below that required to form stoichiometric compounds.

Materials with extrinsic defects are often of great technological relevance; the well-known examples are doped or extrinsic semiconductors. Some of the common forms of extrinsic defects discussed further are:

- substitutional/interstitial impurity
- aliovalent impurity
- charge compensation defect
- non-stoichiometry
- color centers

As noted earlier, the first three cases are considered in this section, and the remaining two in the subsequent sections.

2.3.1 Substitutional/Interstitial Impurity

The presence of a foreign atom, ion, or molecule at an interstitial site or a lattice site (Figure 2.5) leads to this defect. These materials are solid solutions. Some examples are:

- When Mn is introduced in Cu metal, the latter can exhibit spin-glass or bulk magnetic phenomena.
- Doped semiconductors: Si doped with B forms a p-type semiconductor and Si doped with P forms an n-type semiconductor. The dopants occupy the Si sites.

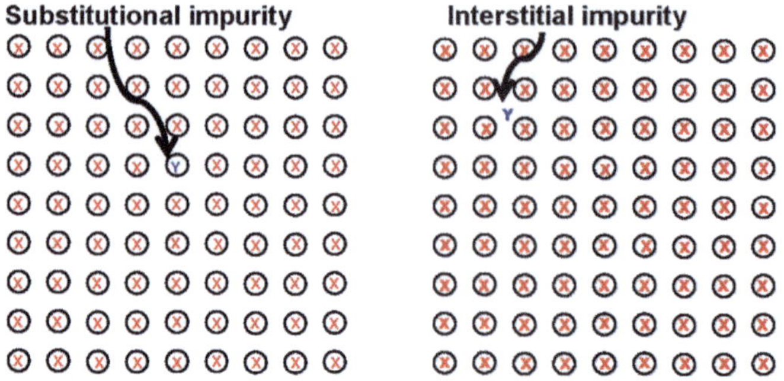

Substitutional impurity   **Interstitial impurity**

Figure 2.5 Schematic diagram of substitutional/interstitial impurities.

- Zn atoms in ZnO is a case of an interstitial impurity; normally, they tend to be labile as they are not strongly bound by electrostatic forces.

2.3.2 Aliovalent Impurity

Aliovalent impurity refers to a dopant with a different oxidation state than that of the ion it replaces in the lattice.

- A well-known example is a Ca^{2+} impurity in a K^+Cl^- crystal. One Ca^{2+} replaces two K^+ in order to maintain overall charge neutrality,
- K^+ ion vacancies form concomitantly in the lattice. This results in enhanced K^+ ion transport in the defective crystal.

2.3.3 Charge Compensation Defect

A typical example is two Ti^{4+} ions in $SrTiO_3$ being substituted by a dication (Co^{2+} or Ni^{2+}) and a hexacation (Mo^{6+} or W^{6+}) so that the total charge of $+8$ is compensated. This provides a convenient and powerful route to form a large number of novel materials.

2.4 Non-stoichiometry

Generally compounds have a simple and well-defined stoichiometry in agreement with the fundamental **law of definite proportions**. However, with many compounds, especially those involving the transition metals that are capable of stable variable valencies, compounds can contain their elements in rather complex and variable ratios.

This occurs especially in the case of early transition metals like Ti and V which have relatively more diffuse d-orbitals which can spread into adjacent unit cells in a crystal; the latter allows the ligand ions from the adjacent unit cells to coordinate to the metal ion facilitating vacancies within the original unit cell.

The idea of non-stoichiometry can be illustrated using an example, $Fe_{1-x}O$.

- Elemental composition analysis of ferrous oxide (FeO) samples rarely show an exact $1:1$ stoichiometric ratio between the elements; usually, they would be deficient in Fe.
- The reason for this is that a small fraction of Fe^{2+} is oxidized to Fe^{3+} and hence slightly less iron is sufficient to compensate for the charge (-2) on oxygen. The formula may be written in detail as $(Fe^{3+})_a(Fe^{2+})_bO$.
- Assume that iron is oxidized to the extent that $\dfrac{Fe^{3+}}{Fe^{2+}} = 0.05$
- Values of a and b can be determined by solving the simultaneous equations:

$$\frac{a}{b} = 0.05$$

$3a + 2b = 2$ (total charge on iron is $+2$, compensating the charge on oxygen)
- Solution of the equations gives, $a = 0.05$, $b = 0.93$. Hence $a + b = 0.98$.
- The formula then works out to be $Fe_{0.98}O$ (in $Fe_{1-x}O$, the value of $x = 0.02$).
- FeO has a rock salt (NaCl) crystal structure; obviously, Fe^{2+} are in the octahedral sites of the O^{2-} lattice. In the non-stoichiometric compound, vacancies of iron sites (Fe:O being less than $1:1$) cluster together, with the Fe^{3+} occupying the tetrahedral sites provided by the cluster.

2.4.1 Classifications

Common examples of non-stoichiometry may be found in metal oxides. Consider a metal compound with the stoichiometric composition, MX. During its formation, if the partial pressure of X_2 (*e.g.*, oxygen) is maintained higher or lower than the equilibrium value required to form the stoichiometric composition ($p_{X_2}(x) > p_{X_2}(0)$ or $p_{X_2}(x) < p_{X_2}(0)$ respectively), non-stoichiometric oxides could form.

The following represent the typical situations that arise. The equations denote the origin of the non-stoichiometric system

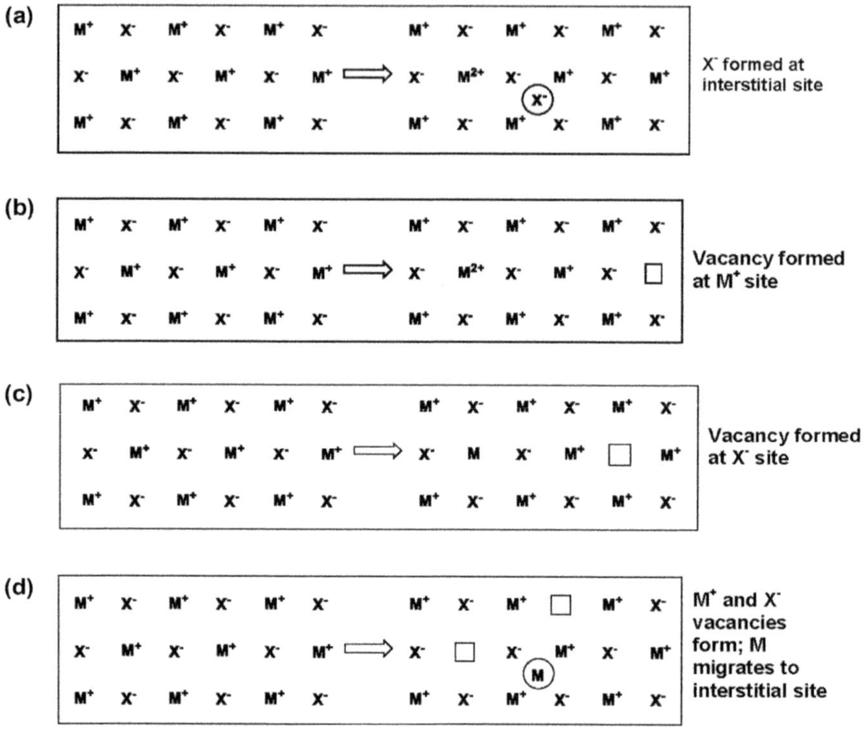

Figure 2.6 Schematic representation of the formation of non-stoichiometry defects in (a) MX_{1+x}, (b) $M_{1-x}X$, (c) MX_{1-x}, and (d) $M_{1+x}X$.

(I and L represent the interstitial and lattice sites respectively); Figure 2.6 shows the defect formation schematically. M^+ and X^- represent metal cations and their counter anions; the actual ions may have charges different from $+1$ and -1.

- The case of $p_{X_2}(x) > p_{X_2}(0)$:
  - MX_{1+x}: $\frac{1}{2}X_2 + e^- \rightarrow X^-(I)$; $M^+ \rightarrow M^{2+} + e^-(L)$ (Figure 2.6(a))

    Example: $UO_2 + \frac{x}{2}O_2 \xrightarrow{1150°C} UO_{2+x} [0 < x < 0.25]$.
    U^{4+} is partially oxidized, and O^{2-} formed at the interstitial sites.
  - $M_{1-x}X$: $\frac{1}{2}X_2 + e^- \rightarrow X^-(L)$; $M^+ \rightarrow M^{2+} + e^-(L)$ (Figure 2.6(b))

    Example: $\left(1 - \frac{x}{2}\right)Cu_2O + \frac{x}{4}O_2 \rightarrow Cu_{2-x}O$
    Cu^+ is partially oxidized, and vacancies appear at Cu^+ sites.
- The case of $p_{X_2}(x) < p_{X_2}(0)$:
  - MX_{1-x}: $X^- \rightarrow \frac{1}{2}X_2 + e^-(L)$; $M^+ + e^- \rightarrow M(L)$ (Figure 2.6(c))

Example: TiO_{1-x}

Ti^{2+} is partially reduced and O^{2-} vacancies form. Titanium oxide is known to form a large range of non-stoichiometric compounds, $TiO_{0.64}$ to $TiO_{1.27}$.

○ $M_{1+x}X$: $X^- \rightarrow \dfrac{1}{2}X_2 + e^-(L)$; $M^+ + e^- \rightarrow M(I)$ (Figure 2.6(d))

Example: $(1+x)ZnO \xrightarrow{\Delta} Zn_{1+x}O + \dfrac{x}{2}O_2$; ZnO is white and $Zn_{1+x}O$ yellow. Zn^{2+} is partially reduced to Zn^{1+} and O^{2-} oxidized with O_2 boiling out.[1]

The equilibria associated with the formation of defects and knowledge of the defect concentration can be used to estimate the relevant thermodynamic parameters. Conversely, experimental determination of the thermodynamic parameters can lead to the determination of the defect concentration.

2.4.2 More Examples of Non-stoichiometric Compounds

Some examples of non-stoichiometric materials with special structural or functional attributes are described below.

- Several non-stoichiometric phases $NiTe_x$ form between the two stoichiometric extremes, NiTe (NiAs structure) and $NiTe_2$ (CdI_2 structure) (Figure 2.7).
- Praseodymium oxide has a range of non-stoichiometric compositions, PrO_{2-x} $(0 < x < 0.25)$ at 1000 °C. At lower temperatures, **infinitely adaptable structures** with the formula Pr_nO_{2n-2} $(n = 4, 7, 9, 10, 11, 12, \infty)$ are observed. The series can be

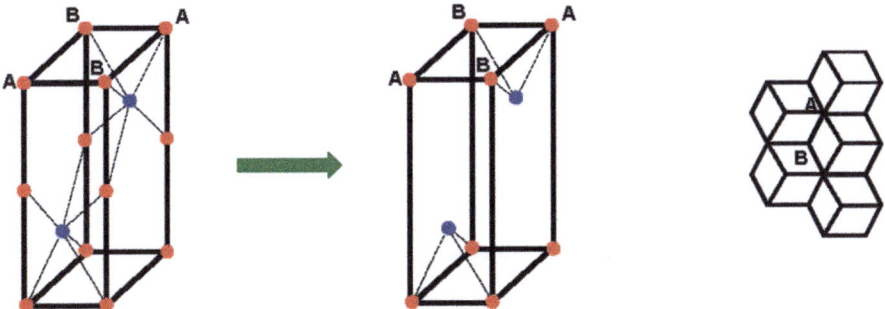

Figure 2.7 NiTe (NiAs structure, Ni (red), As (blue)) and $NiTe_2$ (CdI_2 structure, Cd (red), I (blue)); views of the 2-D lattice of Ni (in NiAs) and Cd (in CdI_2) are shown along with the two types of lattice sites, A and B which are also marked on the 3-D structures.

visualized as ranging from Pr_2O_3 ($n=4$) to PrO_2 ($n=\infty$). PrO_2 has the fluorite structure with Pr forming an fcc lattice and O occupying all the T_d sites (note that a conventional unit cell with 4 atoms has 8 T_d and 4 O_h sites). Pr_2O_3 has the C-type rare earth M_2O_3 structure in which $\frac{3}{4}$ of the T_d sites of the fcc lattice of Pr are occupied by O. The O vacancies organize to form superstructures with unit cell lengths that are multiples of the original lattice value.

- The high temperature superconductors, for example, $YBa_2Cu_3O_{7-x}$ form another important class of non-stoichiometric compounds. When $x=0.5$, all copper are in the $+2$ state. When x decreases (*i.e.*, O content increases), holes are formed in the Cu–O layer (equivalent to the formation of Cu^{3+}); when $x=0$, there are two Cu^{2+} and one Cu^{3+} ion. Structural transformations accompany the variation of x. The material is superconducting only for $x<0.5$.

- Tungsten bronzes are derived from WO_3. Insertion of alkali metals, M (Na, K, Rb, or Cs) gives rise to the non-stoichiometric systems, M_xWO_3. The color and electrical properties are dependent on x. For example, in Na_xWO_3 the color ranges from golden to red, orange, purple, and blue–black as x varies from 0.9 to 0.3; electrical conductivity decreases with x. WO_3 has an ReO_3 structure (Figure 2.8); in Na_xWO_3, Na sits at the body center of the cubic unit cell with W at the vertices and O at the edge centers. The body center has fractional occupation, x; when $x=1$, a perovskite structure is obtained.

- An interesting example of non-stoichiometry in organic solids is the charge transfer (donor–acceptor) complex $(M_2P)_{1-x}(P)_xTCNQ$.

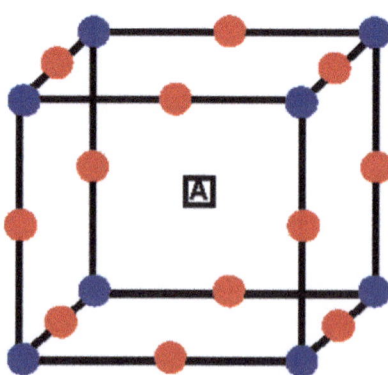

Figure 2.8   Unit cell of WO_3 (ReO$_3$ structure, Re (blue), O (red)); the body-center site, A is occupied by the alkali metal (fractional occupation, x of Na in Na_xWO_3).

P = phenazine, $M_2P = N,N$-dimethyl phenazine (a strong π-electron donor), and TCNQ = tetracyanoquinodimethane (a strong π-electron acceptor). $x = 0$ corresponds to $M_2P^+TCNQ^-$, a semi-conductor. When $x \neq 0$, P occupies the sites of M_2P, but remains neutral as it is a poor donor; $(M_2P^+)_{1-x}(P)_x TCNQ^{(1-x)-}$ has partial ionicity on TCNQ leading to metallic behavior.

- The bromide salt of tetrathiafulvalene, $TTF(Br)_x$ is another electrically conducting organic solid, thanks to the partial oxidation of TTF.

2.5 Color Centers

A typical observation when crystals of alkali metal halides are heated in the presence of a metal vapor is that they acquire a color; the color depends on the crystal and not on the type of metal vapor. The color is attributed to F-centers (*farbe* = color in German). These arise due to an anion vacancy (point defect) in the crystal which traps an electron; if the energy levels of the electron confined in the vacancy site are such that visible light of specific wavelengths can be absorbed, the crystal becomes colored.

The phenomenon can be explained using the NaCl crystal. The reaction that occurs when NaCl is heated in a metal (M) vapor can be represented as follows:

$NaCl(s) + M(g) \rightarrow (NaCl)^- M^+(s)$ [M = Na, K, *etc.*] (Figure 2.9(a))

In crystals of increasing unit cell length, the energy gaps are smaller (recall, the particle-in-a-box model) and wavelength of the light absorbed increases (Table 2.1, Figure 2.9(b)).

- Combinations of two or three F-centers lead to defects called M- and R-centers.
- The unpaired electrons in color centers can be studied using electron paramagnetic resonance.

2.6 Characterization of Defects

The unique features of the various kinds of defects, plane, line, and point, discussed in the previous sections can be used for their experimental characterization. The plane and line defects are often identified using electron microscopy techniques which allow a direct visualization of the defect structure. High resolution microscopy can provide atomic level details facilitating the observation of defect

(a)

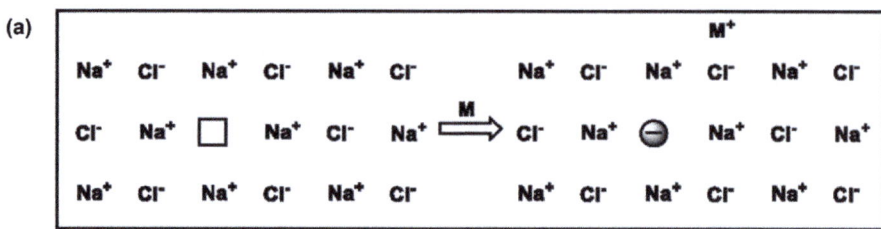

(b)

Figure 2.9 Schematic figure showing (a) the formation of an F-center in NaCl, and (b) the color variation due to F-centers in NaCl, KCl, KBr crystals. Photographs of real crystals can be seen at ref. 2.

Table 2.1 Correlation between unit cell size and F-center light absorption.

Crystal	Unit cell length (Å)	Light absorption	
		Energy (eV)	Wavelength (nm)
NaCl	5.64	2.67	464
KCl	6.29	2.20	564
RbCl	6.58	1.97	629
KBr	6.54	2.00	620

boundaries. Burger's vector noted in Section 2.1 denotes a characteristic signature to identify the defect location. The impact of plane and line defects on the mechanical and structural properties of materials provide an indirect assessment of their presence. X-ray diffraction can also be used to investigate defects, by monitoring the deviations from the diffraction pattern expected in a defect-free crystal. Formation of superstructures in crystals (Section 2.4.2) with high defect concentration and their ordering, can be detected through X-ray diffraction.

Identification of point defects can be carried out through a variety of experimental observations depending on the specific nature of the point defect. For example, as shown in Section 2.2, some defects lead to a decrease in the density of the crystals whereas others do not. Therefore a comparison of the experimentally determined density with that expected theoretically based on the perfect crystalline structure can provide a clue to the presence of defects like the

Schottky type. Several of the point defects and non-stoichiometry cases can be characterized using their impact on the solid state properties such as electrical conductivity (ionic conductivity in particular), color and spectroscopic properties and magnetic behavior. Specific cases such as tungsten bronzes, F-centers, *etc.*, discussed in the earlier sections provide the rationale behind such characterization approaches.

References

1. C. B. Carter and M. G. Norton, *Ceramic Materials: Science and Engineering*, Springer, 2007, p.188.
2. https://chemistry.beloit.edu/edetc/background/F_center/index.htm

3 Thermal Properties

3.1 Lattice Vibrations – Phonon Dispersion

A crystal can be visualized as a giant molecule with the atoms in the basis organized on the lattice with its relevant symmetry elements. Description of the collective oscillations of the n atoms in a molecule in terms of the $(3n - 6)$ vibrational modes, can be extended to the case of crystals as well; as n is very large, it can be considered as $3n$ modes. The translational symmetry allows these vibrational modes to be described utilizing the periodic structure. Collective vibrations involving all particles in a lattice, the harmonic vibrational modes of the lattice, are called **phonons**. Important aspects of phonons are:

- All atoms in a particular mode oscillate with the same frequency.
- Energy of the phonons are quantized.

A 1-D lattice with a monatomic basis, with the unit cell length a, can be used as a simple example to illustrate the idea of phonons.

- Figure 3.1 shows two modes of collective vibrations; as the vibrations are periodic, they can be represented as waves with wavelengths $4a$ and $2a$ respectively.
- Note that the smallest wavelength possible, $\lambda_{min} = 2a$ and the largest, $\lambda_{max} = Na = L$ (N = number of lattice sites and L = length of the crystal).
- ω_{max}(max. angular frequency) $= 2\pi f_{max}$(max. frequency) $= \dfrac{2\pi v}{\lambda_{min}} = \dfrac{\pi v}{a}$.

Core Concepts for a Course on Materials Chemistry
By T. P. Radhakrishnan
© T. P. Radhakrishnan 2023
Published by the Royal Society of Chemistry, www.rsc.org

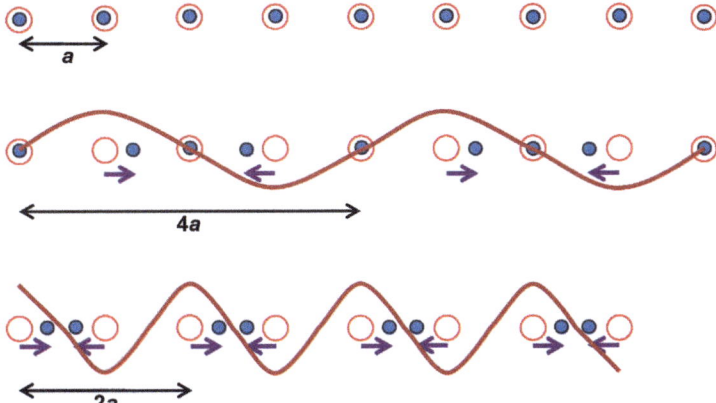

Figure 3.1 1-D lattice (red open circle) with unit cell length *a*, and a monatomic basis (blue filled circle); atomic displacements in the collective vibrations with wavelengths 4*a* and 2*a*.

- v = velocity of wave propagation in the crystal.
- v and a depend on the material. For Cu metal, the velocity of sound propagation (*via* phonons) ~ 3550 m s^{-1} and $a \sim 3.6$ Å; ω_{max} works out to be 3×10^{13} radians s^{-1}.

In general, a wave can be represented by the **wave vector**, $\vec{k} = \dfrac{2\pi}{\lambda} \hat{i}$ (see Section 1.4.6). Some relevant aspects of the wave vector are the following:

- $|\vec{k}| = k = \dfrac{2\pi}{\lambda} = \dfrac{2\pi \cdot f}{\lambda \cdot f} = \dfrac{\omega}{v}$; [$f$ = frequency; v = velocity]

- For the phonons discussed above, $k_{max} = \dfrac{\pi}{a}$; $k_{min} = \dfrac{2\pi}{L}$ ($k_{min} \to 0$ in the large L limit).

- The relation, $\omega = kv$ shows that the phonon frequency is linearly related to the phonon wave vector; v (called the phase velocity) is the proportionality constant. It will be seen below that this is not true for phonons at all frequencies.

3.1.1 Phonon Dispersion Curve

Consider a 1-D lattice with a monatomic basis and only nearest neighbor interactions, represented by the force constant, C.

- The displacement of the atom at the nth lattice point at an instant in time, t during a vibrational motion is denoted as $u_n(t)$;

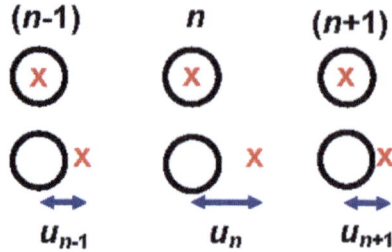

Figure 3.2 Schematic representation of the relative displacement of atoms, u_n at site n, etc.

fixing an arbitrary time, the displacements can be represented simply as u_n (Figure 3.2).

- If the vibrations are considered to be harmonic, the potential energy, E associated with the relative displacements of the neighbors of the atom at position, n is:

$$E = \frac{1}{2}C\left[(u_n - u_{n-1})^2 + (u_{n+1} - u_n)^2\right].$$

- Taking the mass of the atoms as M, the equations of motion can be formulated. The force acting on the atom can be written as mass×acceleration, or the negative derivative of E with respect to displacement. Equating the two,

$$M\left(\frac{\partial^2 u_n}{\partial t^2}\right) = -\frac{\partial E}{\partial u_n} = -C[2u_n - u_{n-1} - u_{n+1}].$$

- The solution of the above equation will be of the form, $u_n(t) \propto e^{i(kna-\omega t)}$ ($a =$ unit cell length) representing the sinusoidal oscillations in space and time.
- Boundary effects in the lattice with N sites are avoided by imposing the Born–von Karman boundary condition,

    $u_{N+1} = u_1$

    This condition implies that, $e^{i(k[N+1]a-\omega t)} = e^{i(ka-\omega t)} \Rightarrow e^{ikNa} = 1$

    $\Rightarrow kNa = 2n\pi$ ($n = 0, \pm 1, \pm 2 \ldots$) $\Rightarrow k = \dfrac{2\pi}{a}\dfrac{n}{N} = \dfrac{2\pi}{L}n$ (quantization of k)

    $\Rightarrow \lambda = \dfrac{Na}{n} = \dfrac{L}{n}$

- Using the solution suggested above, the differential equation gives:

$$-M\omega^2 e^{i(kna-\omega t)} = -C[2 - e^{-ika} - e^{+ika}]e^{i(kna-\omega t)}$$

$$= -2C[1 - \cos(ka)]e^{i(kna-\omega t)}$$

$$\Rightarrow \omega = \sqrt{\frac{2C}{M}[1 - \cos(ka)]}$$

- Writing the frequency explicitly as a function of the wave vector, the phonon dispersion relation is: $\omega(k) = 2\sqrt{\dfrac{C}{M}}\left|\sin\dfrac{1}{2}ka\right|$

- The plot of $\omega(k)$ *versus* k is shown in Figure 3.3.

- The region, $-\dfrac{\pi}{a} < k \leq \dfrac{\pi}{a}$ is called the Brillouin zone. The dispersion curve points repeat for $k' = k \pm n\dfrac{2\pi}{a}$. This can be seen from:

$$\left|\sin\frac{1}{2}\left(k \pm n\frac{2\pi}{a}\right)a\right| = \left|\sin\left(\frac{1}{2}ka \pm n\pi\right)\right| = \left|\sin\frac{1}{2}ka\right|$$

- The above observation can be understood as follows. The allowed values of k with $-\dfrac{\pi}{a} < k \leq \dfrac{\pi}{a}$ is N, the number of vibrational modes of a 1-D chain. Extending k beyond this interval gives exactly the same displacements as those found in the Brillouin zone and does not produce any new vibrational modes. $k=0$ corresponds to the translation of the chain and the corresponding frequency, $\omega = 0$, as seen from the dispersion relation.

- At the $\lim\limits_{k\to 0}\dfrac{\sin\dfrac{1}{2}ka}{\dfrac{1}{2}ka} = 1$, and $\omega(k) = a\sqrt{\dfrac{C}{M}}|k|$. Frequency varies linearly with the wave vector. The dispersion curve in Figure 3.2 is nearly linear at low values of k. These modes are called the acoustic modes. The proportionality constant for $\omega(k) \propto k$ is the sound velocity; this may be compared with the case of photons for which $\omega = c_L k$ where c_L is the velocity of light.

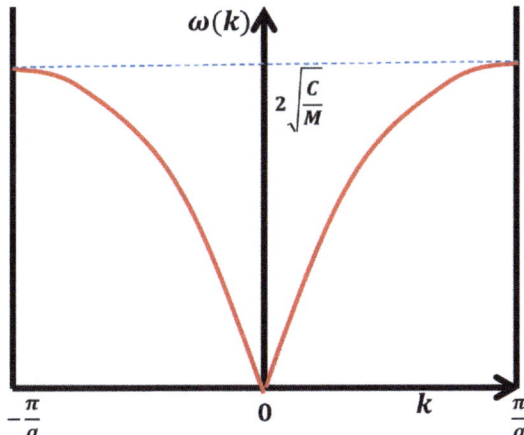

Figure 3.3 Phonon dispersion curve for a 1-D lattice (unit cell length $=a$) with a monatomic basis (atom mass $=M$, force constant for near neighbor interaction $=C$) within the first Brillouin zone.

- When λ is short and close to $2a$ (k is large and close to the Brillouin zone boundary), ω becomes less than kv. Another way to describe this is to write, $\omega = kv$, where $v = v_0 \dfrac{\sin \frac{1}{2}ka}{\frac{1}{2}ka}$ with $v_0 = a\sqrt{\dfrac{C}{M}}$; this means that v depends on k.

3.1.2   Lattice With Multi-atom Basis and Vibrations in 3-D

Consider a 1-D lattice with a diatomic basis having atoms with masses M_1 and M_2. Typical vibrational modes are shown in Figure 3.4. Two classes of phonons can be observed, one in which there are no vibrations within the basis (atoms in the basis oscillate in phase) and the other in which there are vibrations within the basis. The former are the acoustic modes (involved in sound propagation), while the latter are the optical modes (involved in interactions with light). Figure 3.4 also shows the dispersion curves.

In a linear lattice, if the atoms can vibrate in three directions, not only longitudinally (parallel to the chain axis), but also in the two transverse directions (perpendicular to the chain axis), then there are three modes, one longitudinal and two transverse; these give rise to three branches in the phonon dispersion curves (Figure 3.5).

- For a 3-D lattice with a p-atom basis, there are 3p branches: 3 acoustic, 3(p − 1) optical.
- The transverse acoustic modes have lower frequency than the longitudinal ones, as the former involve bending force constants while the latter involve stretching force constants; in physical systems, the bending mode frequencies are lower than the stretching ones.

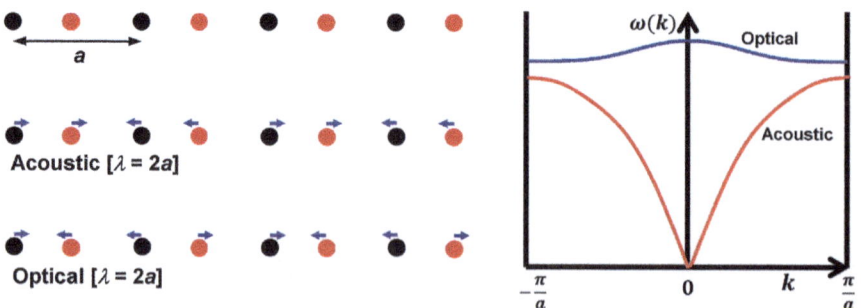

Figure 3.4   1-D lattice with a diatomic basis; the red and black filled circles represent the atoms with different masses. The unit cell length, *a* and the atom displacements for acoustic and optical modes with $k = \pi/a$ are shown. Phonon dispersion curves showing the acoustic and optical branches.

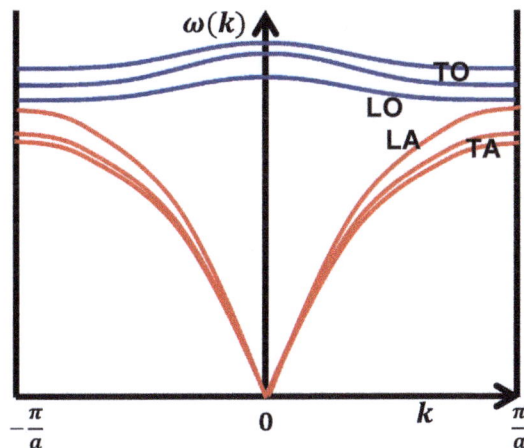

Figure 3.5   Phonon dispersion curves for a linear lattice (unit cell length $=a$) with diatomic basis and vibrations in three directions; LA $=$ longitudinal acoustic; TA $=$ transverse acoustic; LO $=$ longitudinal optical; TO $=$ transverse optical.

3.1.3   Experimental Determination of Phonon Dispersion Curve

Phonons do not carry any physical momentum, as they have relative coordinates, and the center of mass of the crystal does not move during the oscillations $[M\dfrac{\partial}{\partial t}\sum u_n = 0]$. However, they can interact with photons, neutrons, electrons, *etc.*, as if they had the momentum, $\hbar\vec{k}$ (sometimes called **crystal momentum**). Elastic scattering (with conservation of total momentum) of X-ray photons by a crystal is governed by the condition (selection rule), $k - k' = K$, where K is a reciprocal lattice vector. The phonon dispersion curve is often determined using neutron scattering experiments

- Extending the equation above, the momentum conservation for the inelastic scattering (with exchange of energy) of neutrons can be written in terms of the wave vectors of the incident and scattered neutrons (k_n, k'_n), the phonon created or annihilated (k_p) and the reciprocal lattice vector (K):

  $k_n - k'_n = K \pm k_p.$

- The equation shows that momentum is exchanged between the neutrons and the crystal; the crystal gains or loses the phonon momentum.
- The corresponding energy conservation equation is:

$$\frac{\hbar|k_n|^2}{2M_n} = \frac{\hbar|k'_n|^2}{2M_n} \pm \hbar\omega(k_p).$$

- The phonon dispersion curve can be determined by experimentally evaluating the energy loss or gain of the scattered neutrons as a function of the scattering direction, $k_n - k'_n$.

3.2 Lattice Heat Capacity

Heat capacity is a fundamentally important characteristic of matter. It determines the various thermal properties of materials.

3.2.1 Quantization of Lattice Vibrations – Phonons

Each normal mode has energy and momentum defined by its frequency, ω and wave vector, k. Energy of a harmonic oscillator in state n, $E_n = \left(n + \frac{1}{2}\right)\hbar\omega$ (Figure 3.6(a); the energy gap indicated is $\hbar\omega$). This is the quantization of energy for phonons. One can say that the system has n phonons with wave vector, k, if the vibrational quantum number in the mode k is n.

Phonon energy distribution at temperature, T is given by the Bose–Einstein distribution: $f(\omega) = \dfrac{1}{e^{\hbar\omega/k_B T} - 1}$. This is essentially the probability of excitation of a mode of frequency, ω at temperature T. In other words, if there are N oscillators, $Nf(\omega)$ oscillators possessing energy, $\hbar\omega$ are excited at temperature T. Predominantly, modes possessing $\hbar\omega \leq k_B T$ are excited. The Bose–Einstein distribution function is shown in Figure 3.6(b); the excitation of modes with different frequencies at a temperature T are also shown schematically.

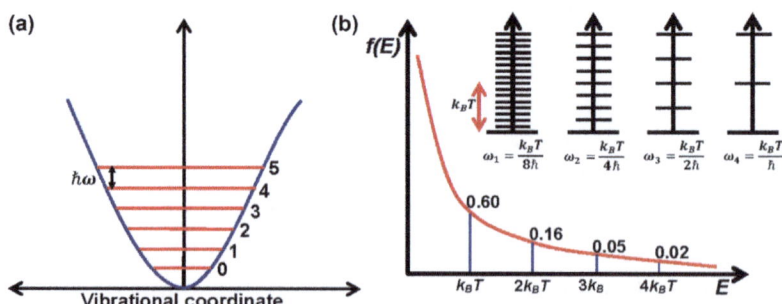

Figure 3.6   (a) Quantized energy levels of a phonon with frequency ω; the quantum numbers are shown. (b) Bose–Einstein distribution function for phonons with different fundamental frequencies at temperature T are shown schematically. Drawn schematically based on Figures 4.3 and 4.4 in ref. 1.

- It is seen from Figure 3.5(b) that for a temperature, T, ω_1 mode is excited to the eighth level (ω_1 mode generates eight phonons of energy, $\hbar\omega = \dfrac{k_B T}{8}$), ω_2 to the fourth level, *etc.* Modes with frequency greater than ω_4 are not excited appreciably.
- The energy of N oscillators is given by:

$$U = N\hbar\omega f(\omega) = \frac{N\hbar\omega}{e^{\hbar\omega/k_B T} - 1}.$$

3.2.2 Einstein Model for Heat Capacity

Einstein made the assumption that oscillators in the solid possess a single frequency, ω (now called the Einstein frequency, ω_E) contributing to the lattice heat capacity. Based on the equation above, the energy for N oscillators with displacements in 3-D can be written as:

$$U = \frac{3N\hbar\omega}{e^{\hbar\omega/k_B T} - 1}$$

Using the derivative,

$$\frac{\partial \left(e^{\hbar\omega/k_B T} - 1 \right)^{-1}}{\partial T} = \frac{-1}{\left(e^{\hbar\omega/k_B T} - 1 \right)^2} \times e^{\hbar\omega/k_B T} \times \frac{-\hbar\omega}{k_B T^2},$$

the heat capacity is given by:

$$C_V = \left(\frac{\partial U}{\partial T} \right)_V = 3Nk_B \left(\frac{\hbar\omega}{k_B T} \right)^2 \frac{e^{\hbar\omega/k_B T}}{\left(e^{\hbar\omega/k_B T} - 1 \right)^2}$$

In the high-temperature limit, $\hbar\omega \ll k_B T$, so that

$$e^{\hbar\omega/k_B T} \simeq 1 + \frac{\hbar\omega}{k_B T}. \quad \therefore \quad U = \frac{3N\hbar\omega}{\hbar\omega/k_B T} = 3Nk_B T$$

- Heat capacity, $C_V = 3Nk_B$; if N = Avogadro number, the molar heat capacity = $3R$.
- This is the **Dulong–Petit law**: molar heat capacity of monatomic solids, $C = 3R \sim 25$ J K^{-1} mol^{-1}.

In the low-temperature limit, $\hbar\omega \gg k_B T$, $e^{\hbar\omega/k_B T} \gg 1$. $\therefore U = \dfrac{3N\hbar\omega}{e^{\hbar\omega/k_B T}}$.

- Heat capacity can now be written as: $C_V = 3Nk_B \left(\dfrac{\hbar\omega}{k_B T} \right)^2 e^{-\hbar\omega/k_B T}$

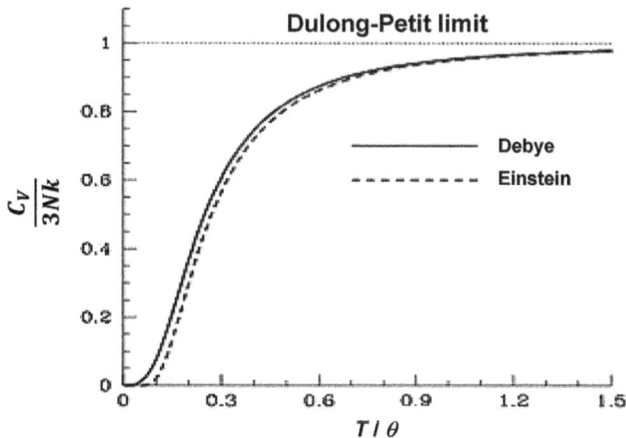

Figure 3.7 Temperature dependence of heat capacity of solids ($\theta =$ Debye or Einstein temperature). Adapted from ref. 1.

- This shows that the heat capacity at low temperatures decreases exponentially (the square term has a weaker influence than the exponential factor) with the temperature.
- This conclusion contradicts the experimental observation at low temperatures, that the heat capacity varies as, $C_V \sim T^3$. Figure 3.7 shows the variation of heat capacity with reduced temperature following the Einstein model and Debye's T^3 law.

3.2.3 Debye Model for Heat Capacity

Debye took into consideration the possibility that modes with a range of frequencies contribute to the lattice heat capacity. The following assumptions were made:

- Modes with **frequency** up to a **cut-off value** would be included (the cut-off is now called the Debye frequency, ω_D). ω_D is the frequency of the highest energy oscillator after cut-off.
- **Linear relation** between frequency and wave vector (this is justified in the low frequency region as discussed in Section 3.1.1).

Considering a distribution of frequencies given by the **density of states, $\mathfrak{D}(\omega) = \dfrac{\partial N(\omega)}{\partial \omega}$** (the number of oscillators with frequency in the range ω to $\omega + d\omega$), the total energy assuming oscillations in 3-D can be written as:

$$U = 3 \int_0^{\omega_D} \mathfrak{D}(\omega) \cdot \hbar\omega \cdot f(\omega) \cdot d\omega = 3 \int_0^{\omega_D} \frac{\mathfrak{D}(\omega) \cdot \hbar\omega}{e^{\hbar\omega/k_B T} - 1} d\omega.$$

The expression for the density of states, $\mathfrak{D}(\omega)$ can be determined first:

- The number of oscillators with wave vectors up to a value k is obtained by counting the number of points in the reciprocal lattice contained in a sphere of radius k. The number of reciprocal lattice points should be equal to the number of vibrational degrees of freedom.
- The reciprocal lattice points are given by the quantization condition, $k = \dfrac{2\pi}{L} n$ (Section 3.1.1). So each point can be assumed to occupy a volume, $\left(\dfrac{2\pi}{L}\right)^3$.
- The number of k-points, $N(k) = \dfrac{\frac{4}{3}\pi k^3}{\left(\dfrac{2\pi}{L}\right)^3} = \dfrac{Vk^3}{6\pi^2}$ ($V =$ volume of the crystal).
- Making the linear approximation $\omega = kv$, $N(\omega) = \dfrac{V\omega^3}{6\pi^2 v^3}$.
- Therefore, $\mathfrak{D}(\omega) = \dfrac{\partial N(\omega)}{\partial \omega} = \dfrac{V\omega^2}{2\pi^2 v^3}$.

The total energy can now be written as: $U = \dfrac{3V\hbar}{2\pi^2 v^3} \displaystyle\int_0^{\omega_D} \dfrac{\omega^3}{e^{\hbar\omega/k_B T} - 1} d\omega$

- Define: $\dfrac{\hbar\omega}{k_B T} = x$; $\dfrac{\hbar\omega_D}{k_B T} = x_D$; $\dfrac{\hbar\omega_D}{k_B} = \theta$ (the Debye temperature).
- $N(\omega) = \dfrac{V\omega^3}{6\pi^2 v^3} \Rightarrow N = \dfrac{V\omega_D^3}{6\pi^2 v^3}$ (all of the N oscillators have $\omega \leq \omega_D$)

$$\Rightarrow \omega_D^3 = \dfrac{6\pi^2 v^3 N}{V}.$$

The total energy can now be written as:

$$U = \dfrac{3V\hbar}{2\pi^2 v^3} \left(\dfrac{k_B T}{\hbar}\right)^4 \int_0^{x_D} \dfrac{x^3}{e^x - 1} dx$$

$$= \dfrac{3Vk_B}{2\pi^2 v^3} \left(\dfrac{k_B}{\hbar}\right)^3 T^4 \int_0^{x_D} \dfrac{x^3}{e^x - 1} dx$$

$$= 9k_B \left(\dfrac{V}{6\pi^2 v^3}\right) \left(\dfrac{k_B}{\hbar}\right)^3 T^4 \int_0^{x_D} \dfrac{x^3}{e^x - 1} dx$$

$$= 9Nk_B \left(\dfrac{k_B}{\hbar\omega_D}\right)^3 T^4 \int_0^{x_D} \dfrac{x^3}{e^x - 1} dx$$

$$U = 9Nk_B T \left(\dfrac{T}{\theta}\right)^3 \int_0^{x_D} \dfrac{x^3}{e^x - 1} dx.$$

In the high temperature limit, $\hbar\omega \ll k_B T$, x is small and $e^x \simeq 1 + x$. Therefore,

- $U = 9Nk_B T \left(\dfrac{T}{\theta}\right)^3 \displaystyle\int_0^{x_D} x^2\,dx = 9Nk_B T \left(\dfrac{T}{\theta}\right)^3 \dfrac{x_D^3}{3} = 3Nk_B T$
- Therefore, the molar heat capacity, $C_V = 3R$. (Dulong–Petit).

In the low temperature limit, as $T \to 0$, $x_D \to \infty$.

- $U = 9Nk_B T \left(\dfrac{T}{\theta}\right)^3 \displaystyle\int_0^\infty \dfrac{x^3}{e^x - 1}\,dx$
- $\displaystyle\int_0^\infty \dfrac{x^3}{e^x - 1}\,dx = \sum_s \int_0^\infty x^3 e^{-sx}\,dx = \sum_s \dfrac{6}{s^4} = \dfrac{\pi^4}{15}$

[The above result is from:

$$\dfrac{1}{e^x - 1} = \dfrac{e^{-x}}{1 - e^{-x}} = e^{-x}(1 - e^{-x})^{-1} = e^{-x}(1 + e^{-x} + e^{-2x} + \ldots) = \sum_s e^{-sx},$$

with standard integral: $\displaystyle\int_0^\infty x^n e^{-ax}\,dx = \dfrac{n!}{a^{n+1}}$, and standard summation:

$$\sum_s \dfrac{1}{s^4} = \dfrac{\pi^4}{90}.]$$

- Therefore, $U = \dfrac{3\pi^4}{5} Nk_B \dfrac{T^4}{\theta^3}$
- $C_V = \dfrac{12\pi^4}{5} Nk_B \left(\dfrac{T}{\theta}\right)^3 = 234 Nk_B \left(\dfrac{T}{\theta}\right)^3$. This is known as **Debye's T^3 law.**

The T^3 dependence of C_V can be understood using a qualitative argument as follows.

- At any low temperature, T, the fraction of k space occupied by the excited modes is:
$$\left(\dfrac{k_T}{k_D}\right)^3 = \left(\dfrac{\omega_T}{\omega_D}\right)^3 = \left(\dfrac{T}{\theta}\right)^3.$$
- Thus, there are of the order of $N \left(\dfrac{T}{\theta}\right)^3$ excited modes. As each has energy, $k_B T$, $U \sim Nk_B T \left(\dfrac{T}{\theta}\right)^3$.
- Therefore the heat capacity, $C_V \propto 4 Nk_B \left(\dfrac{T}{\theta}\right)^3$.

The Einstein model is appropriate for optical phonons and the Debye model is appropriate for acoustic phonons. Since at low temperatures only the low energy phonons are excited, the Debye model provides the correct low temperature behavior of the specific heat.

Note finally that the above expressions for heat capacity apply to insulating materials. In the case of metals, there will be an additional contribution to heat capacity from the free electrons; this will be discussed in Chapter 4.

3.3 Thermal Expansion

If the vibrations of atoms in a solid were perfectly harmonic, there would not be any thermal expansion, since the equilibrium bond distances will be the same for all vibrationally excited states. Expansion is a consequence of anharmonicity of the vibrations. Potential energy and hence force associated with displacement, x can be approximated (for relatively small values of x) by:

$$U(x) = \frac{C}{2}x^2 - \frac{D}{3}x^3 \Rightarrow F(x) = -\frac{\partial U(x)}{\partial x} = -Cx + Dx^2,$$

where C and D are the harmonic and anharmonic force constants respectively. How the expansion comes about due to anharmonicity of the vibration is shown schematically in Figure 3.8.

The average value of force associated with displacement, $\bar{x}$ is given by: $\bar{F}(x) = -C\bar{x} + D\overline{x^2}$

- mean force for free vibration is zero; therefore,

$$-C\bar{x} + D\overline{x^2} = 0 \Rightarrow \bar{x} = \frac{D}{C}\overline{x^2}$$

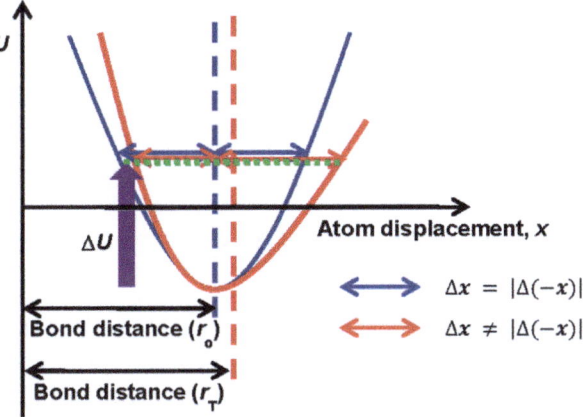

Figure 3.8 Energy variation with displacement for a harmonic (blue curve) and an anharmonic (red curve) potential; for an energy increase, ΔU, the mean displacement is zero for the former (blue double arrows), but non-zero for the latter (red double arrows) where the bond distance increases from r_0 to r_T. Drawn schematically based on Figure 4.7 in ref. 1.

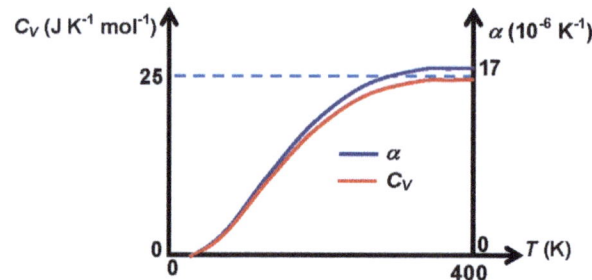

Figure 3.9 Temperature dependence of the heat capacity (C_V) and linear expansion coefficient (α) of copper. Drawn schematically based on Figure 4.8 in ref. 2.

- mean potential energy, $\bar{U}(x) \simeq \dfrac{C}{2}\overline{x^2}$ (to second order, as $\overline{x^3}$ is negligible).
- therefore, $\overline{x^2} = \dfrac{2\bar{U}(x)}{C} \Rightarrow \bar{x} = \dfrac{2D}{C^2}\bar{U}(x)$
- as total average energy, $\bar{E} = \overline{KE} + \overline{PE} = 2\overline{PE}$, $\bar{x} = \dfrac{D}{C^2}\bar{E}$.

If the equilibrium length of the solid is, r_0, the linear expansion coefficient,

$$\alpha = \frac{1}{r_0}\frac{\partial \bar{x}}{\partial T} = \frac{D}{C^2 r_0}\frac{\partial \bar{E}}{\partial T} = \frac{D}{C^2 r_0}C_V.$$

Temperature dependence of α is compared to that of C_V, for a typical case (copper) in Figure 3.9. At high temperatures, C_V is a constant, hence α is also independent of temperature.

For metals, $\alpha = \dfrac{\gamma\chi}{3V}C_V$ where γ is called the Grüneisen constant, χ is the compressibility and V is the atomic volume. The Grüneisen constant takes into account the change in phonon frequencies due to lattice expansion.

3.3.1 Negative Thermal Expansion

Even though most materials expand on heating, several special cases are known to contract when heated. Negative thermal expansion (NTE) can occur due to a variety of mechanisms. The most common ones relate to anisotropic or flexible network structures in the crystal lattice, or phase transitions that lead to a volume contraction.

The directional interactions in liquid water underlie its famous NTE below 3.984 °C. Silicates such as β-eucryptite ($LiAlSiO_4$) are important examples of anisotropic structures exhibiting NTE.

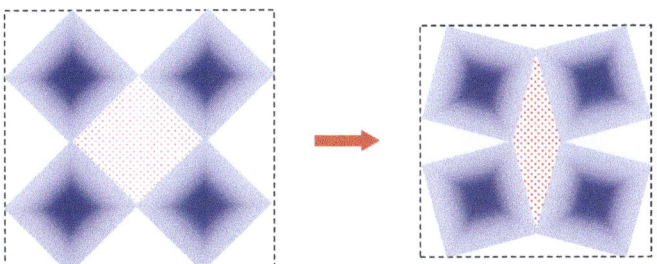

Figure 3.10   Schematic illustration of a vibrational mode in a flexible network of octahedral units (blue), and the contraction of the
volume due to its excitation that leads to reduction of the
open space (represented by the red dots). Drawn schematically
based on Figure 3 in ref. 3.

- Their anisotropy arises due to the strong covalent bonds (Si–O,
  Al–O) and the relatively weaker ionic interactions (Li–O).
- On heating, expansion in the in-plane direction (a-axis) dominated by ionic interactions is overwhelmed by the contraction in
  the out-of-plane direction (c-axis) dominated by the covalent
  bonds leading to overall reduction of volume.

The most extensively studied case of flexible network structures
exhibiting NTE is zirconium tungstate (ZrW_2O_8), showing continuous
contraction on heating from 0.3 to 1050 K.

- It has rigid units of covalently linked ZrO_6 octahedra and WO_4
  tetrahedra; they are corner-linked with one of the corners of the
  latter having no linkage.
- On heating, even though Zr–O and W–O bond distances within
  the rigid units do not change, the weak Zr–O–W linkages allow
  transverse oxygen displacement; these are called rigid unit modes.
- Excitation of these vibrational modes leads to effective reduction of
  the vacant space in the network structure and volume contraction;
  Figure 3.10 illustrates schematically, such a vibrational mode.

A variety of phase transitions including magnetic, ferroelectric,
charge-transfer, and metal-insulator types can lead to NTE in specific
temperature ranges.

Composites of NTE materials with those having normal thermal
expansion behavior can be formed to produce special materials with
negligible or zero thermal expansion; they are extremely useful in
many practical applications such as precision tools; a popular example is that of dental fillings.

3.4 Thermal Conduction

Resistance to the conduction of heat in solids is another consequence of the anharmonicity of the atomic vibrations.

The coefficient of thermal conductivity, κ is defined as the proportionality constant between the flux (flow through unit cross-sectional area in unit time) of thermal energy and the temperature gradient, $J = -\kappa \dfrac{\mathrm{d}T}{\mathrm{d}x}$ (T is the temperature and x is the displacement along the direction of heat flow; note also that the flux is opposite to the temperature gradient, *i.e.*, if T increases in the $+x$ direction, the flux is in the $-x$ direction).

A simple approach to describe thermal conductivity is based on treating phonons as particles that follow the kinetic theory of gases.

- The flux of particles along the $+x$-axis is $\dfrac{1}{2} n \langle |v_x| \rangle$ where n is the number of particles per unit volume (concentration); average of the magnitude of the x-component of the velocity is involved, and equal flux in both directions is assumed.
- When a particle with heat capacity c moves from a region with temperature $T + \Delta T$ to one with temperature T, it transports energy, $c\Delta T$.
- The temperature difference is given by, $\Delta T = \dfrac{\mathrm{d}T}{\mathrm{d}x} l = \dfrac{\mathrm{d}T}{\mathrm{d}x} v_x \tau$, where l is the mean free path and τ is the time between collisions (also called collision time).
- The net flux of energy (considering both $+x$ and $-x$ directions) can now be written as: $J = -n \langle v_x^2 \rangle c\tau \dfrac{\mathrm{d}T}{\mathrm{d}x}$.

  [Note: $+x$ direction: $-\dfrac{n}{2} \langle |v_x| \rangle v_x c\tau \dfrac{\mathrm{d}T}{\mathrm{d}x}$; $-x$ direction: $\dfrac{n}{2} \langle |v_x| \rangle (-v_x) c\tau \dfrac{\mathrm{d}T}{\mathrm{d}x}$]
- Assuming isotropic velocities, $\langle v_x^2 \rangle = \dfrac{1}{3} \langle v^2 \rangle$ (since $v^2 = v_x^2 + v_y^2 + v_z^2$).
- Therefore, $J = -\dfrac{1}{3} n \langle v^2 \rangle c\tau \dfrac{\mathrm{d}T}{\mathrm{d}x} = -\dfrac{1}{3} ncvl \dfrac{\mathrm{d}T}{\mathrm{d}x} = -\dfrac{1}{3} Cvl \dfrac{\mathrm{d}T}{\mathrm{d}x}$

  (C = heat capacity for n particles)
- It directly follows that, $\kappa = \dfrac{1}{3} Cvl$

The mean free path, l is equal to the wavelength of phonons, $\lambda_{\mathrm{ph}} = \dfrac{1}{n_{\mathrm{ph}} \sigma_{\mathrm{ph}}} \propto \dfrac{1}{n_{\mathrm{ph}} D^2}$, where n_{ph} is the phonon concentration, σ_{ph} is

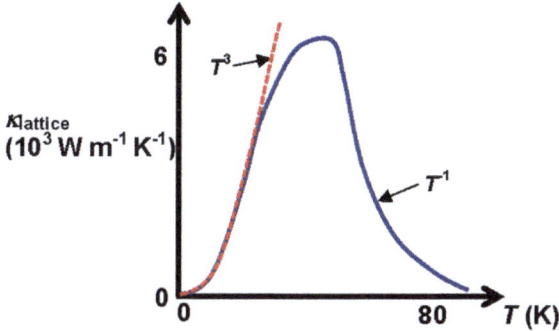

Figure 3.11 Temperature dependence of the thermal conductivity (κ) of synthetic sapphire. Drawn schematically based on Figure 4.9 in ref. 2.

the scattering cross-section which is proportional to the square of the anharmonicity constant (D). If the vibrations are perfectly harmonic, the elastic waves would be non-interacting, and hence will propagate without scattering and hence without thermal resistance.

The thermal conductivity due to phonons, $\kappa_{\text{lattice}} \propto \dfrac{v \cdot C_V}{n_{\text{ph}} D^2}$.

- At low temperatures, n_{ph} is very small. Therefore, λ_{ph} is very large, compared to the crystal dimensions. Hence the temperature dependence is exclusively due to the heat capacity term which is proportional to T^3 (Section 3.2.3). Therefore, $\kappa_{\text{lattice}} \propto T^3$.
- At high temperatures, heat capacity is a constant. The phonon concentration can be shown to be proportional to the temperature: $n_{\text{ph}} = \dfrac{N}{e^{\hbar\omega/k_B T} - 1} \simeq \dfrac{N k_B T}{\hbar\omega}$. Therefore, $\kappa_{\text{lattice}} \propto \dfrac{1}{T}$.
- The complex behavior of κ_{lattice} is seen in the case of synthetic sapphire (Figure 3.11).

The thermal conduction in metals is mediated by the phonons as well as the free electrons, $\kappa = \kappa_{\text{lattice}} + \kappa_{\text{electron}}$; the latter part will be discussed in Chapter 4.

References

1. https://commons.wikimedia.org/wiki/File:DebyeVSEinstein.jpg
2. G. I. Epifanov, *Solid State Physics*, Mir Publishers, Moscow, 1979.
3. K. Takenaka, *Front. Chem.*, 2018, **6**, 267.

4 Electrical Properties

4.1 Free Electron Theory

Free electron theory is one of the earliest attempts to systematically understand the properties of metals. The theory developed by Drude, Sommerfeld, and others provides useful insights into many properties of metals including heat capacity, electrical conduction (Ohm's law), thermal conduction, *etc.*

- However, it fails to explain the basis for the differences between metals and semiconductors.
- The free electron model takes into account the valence electrons of the atoms that constitute the solid. It may be noted that the ion cores (nucleus and core electrons), occupy only about 15% of the volume of the crystal in typical bcc metals.

4.1.1 Free Electron Fermi Gas

The following are the basic assumptions of the free electron theory.

- The total energy of electrons is all kinetic.
- Potential energy is fully neglected. Force between electrons is assumed to be zero (independent electron approximation); similarly forces between the electrons and the ion cores are neglected (free electron approximation).
- The electrons are free to move around, behaving like non-interacting gas molecules in a vessel; however, they are subject

Core Concepts for a Course on Materials Chemistry
By T. P. Radhakrishnan
© T. P. Radhakrishnan 2023
Published by the Royal Society of Chemistry, www.rsc.org

to the Pauli principle. This leads to the description: **free electron Fermi gas.**

A free electron in a crystal can be treated as a particle-in-a-box. Consider first, a 1-D box.

- The Schrödinger equation, considering only the kinetic energy with potential energy $U=0$ is: $-\dfrac{\hbar^2}{2m}\dfrac{\mathrm{d}^2\psi_n}{\mathrm{d}x^2}=\varepsilon_n\psi_n$.
- Imposing the boundary conditions, $\psi_n(0)=\psi_n(L)=0$, solutions for the wave function are: $\psi_n=A\cdot\sin\left(\dfrac{n\pi}{L}x\right)$. ($A$ is the normalization constant, L is the length of the box (crystal), $n\,(=1,2\ldots)$ is the quantum level with the associated wavelength, $\lambda_n=\dfrac{2L}{n}$).
- Note that the solution can also be written as: $\psi_{k_x}=A\cdot\sin(k_x x)$ where $k_x=\dfrac{2\pi}{\lambda}$; the electron wave vector is the quantum number.
- The corresponding energy is: $\varepsilon_n=\dfrac{\hbar^2}{2m}\left(\dfrac{n\pi}{L}\right)^2=\dfrac{n^2h^2}{8mL^2}\left\{\varepsilon_{k_x}=\dfrac{\hbar^2k_x^2}{2m}\right\}$.

The energy levels (free electron orbitals) are filled obeying the Pauli principle. Each level takes two electrons with spin quantum numbers, $m_S=+\frac{1}{2}$ and $m_S=-\frac{1}{2}$.

- If there are a total of N electrons, the top-most filled level has the quantum number, $n_F=\dfrac{N}{2}$. Fermi energy is the energy of this level: $\varepsilon_F=\dfrac{\hbar^2}{2m}\left(\dfrac{n_F\pi}{L}\right)^2=\dfrac{\hbar^2}{2m}\left(\dfrac{N\pi}{2L}\right)^2$.

In 3-D, the Schrödinger equation and solutions can be written as follows:

- $-\dfrac{\hbar^2}{2m}\left(\dfrac{\mathrm{d}^2}{\mathrm{d}x^2}+\dfrac{\mathrm{d}^2}{\mathrm{d}y^2}+\dfrac{\mathrm{d}^2}{\mathrm{d}z^2}\right)\psi_k=-\dfrac{\hbar^2}{2m}\nabla^2\psi_k=\varepsilon_k\psi_k$.
- The solutions are the travelling plane waves, $\psi_k(r)=c\cdot e^{i\vec{k}\cdot\vec{r}}$ where, $k_x=\pm\dfrac{2n\pi}{L}$; $n=0,1,2,\ldots$; similarly, k_y and k_z, and L is the side of the cubic box.
- Energy, $\varepsilon_k=\dfrac{\hbar^2}{2m}\left(k_x^2+k_y^2+k_z^2\right)=\dfrac{\hbar^2k^2}{2m}$.
- In the ground state of the system with N electrons, the occupied levels may be represented as points inside a sphere (in reciprocal

space) of radius, k_F (Fermi wave vector). The energy at the surface, the Fermi energy can be written as: $\varepsilon_F = \dfrac{\hbar^2 k_F^2}{2m}$.

4.1.2 Fermi–Dirac Distribution

The electronic ground state mentioned above occurs at temperature, $T = 0$ K. As T increases, the kinetic energy of the electrons increases and the probability of occupation of the energy levels changes. The probability is given by the Fermi–Dirac distribution:

$$f(\varepsilon) = \frac{1}{e^{(\varepsilon - \mu)/k_B T} + 1}$$

Compare this with the Bose–Einstein distribution discussed in Section 3.2.1; particles with integer spins (Bosons) follow Bose–Einstein statistics, whereas particles with half-odd integer spins (Fermions) follow Fermi–Dirac statistics.

$f(\varepsilon)$ gives the probability that energy level ε is occupied in an ideal electron gas at thermal equilibrium. The energies are referred with respect to the **chemical potential, μ.**

- The chemical potential depends on temperature, and is chosen in such a way that the total number of electrons in the system works out to be the correct value, N.
- At all temperatures, $f(\varepsilon) = \frac{1}{2}$ when $\varepsilon = \mu$.
- Schematic plot of $f(\varepsilon)$ against ε at different temperatures is shown in Figure 4.1.
- It is seen in this example, that the highest energy level occupied at 0 K is 5 eV; this is the Fermi energy, ε_F and $\varepsilon_F = \mu$.

for $\varepsilon < \varepsilon_F, f(\varepsilon) = 1$
for $\varepsilon > \varepsilon_F, f(\varepsilon) = 0$
for $\varepsilon = \varepsilon_F, f(\varepsilon) = 1/2$.

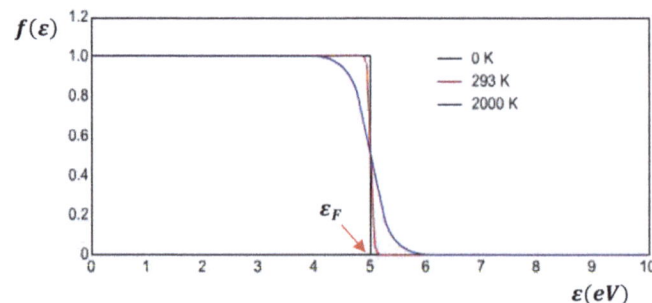

Figure 4.1   Fermi–Dirac distribution at different temperatures. Drawn schematically, based on the plots in ref. 1.

4.1.3   Electronic Density of States

Filling of electrons in energy levels can be described graphically in terms of occupation of the wave vector points in reciprocal space. The filled states are then the points inside a sphere with radius k_F; the Fermi sphere is shown schematically in Figure 4.2(a).

As shown in Section 3.2.3 (for phonons), but including an additional factor of two for the number of spin states per energy level, the density of orbitals per unit energy (density of states, $\mathfrak{D}(\varepsilon)$) for electrons can be derived as follows.

- The number of electrons filled up to level k,

$$N(k) = 2 \left[\frac{\frac{4}{3}\pi k^3}{\left(\frac{2\pi}{L}\right)^3} \right] = \frac{Vk^3}{3\pi^2}.$$

- Since $\varepsilon = \dfrac{\hbar^2 k^2}{2m}$, this equation can be written as,

$$N(\varepsilon) = \frac{V}{3\pi^2} \left(\frac{2m\varepsilon}{\hbar^2} \right)^{3/2}.$$

- $\mathfrak{D}(\varepsilon) = \dfrac{dN(\varepsilon)}{d\varepsilon} = \dfrac{V}{3\pi^2} \dfrac{3}{2} \dfrac{2m}{\hbar^2} \left(\dfrac{2m\varepsilon}{\hbar^2} \right)^{1/2} = \dfrac{Vm}{\pi^2 \hbar^2} \left(\dfrac{2m\varepsilon}{\hbar^2} \right)^{1/2}.$

- Plot of this function (Figure 4.2(b)) shows the filling of orbitals at 0 K up to $\varepsilon = \varepsilon_F$.

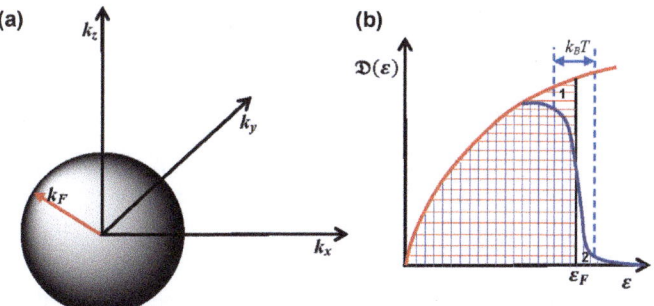

Figure 4.2   (a) Fermi sphere occupied by the system of N free electrons at 0 K. (b) Density of states as a function of the orbital energy (red curve) and the electron filling at 0 K (horizontal red lines). Density of filled orbitals at temperature T ($k_B T \ll \varepsilon_F$) is shown (blue curve) along with the electron filling (vertical blue lines). Drawn schematically based on Figures 4 and 5 in Chapter 6 of ref. 2.

- As the full set of N electrons fill the levels up to the Fermi energy (corresponding to the Fermi sphere with radius, k_F),

$$N = \frac{Vk_F^3}{3\pi^2} \Rightarrow k_F = \left(\frac{3\pi^2 N}{V}\right)^{1/3} \Rightarrow \varepsilon_F = \frac{\hbar^2}{2m}\left(\frac{3\pi^2 N}{V}\right)^{2/3}$$

- The density of filled orbitals at temperature, T (with $k_B T \ll \varepsilon_F$), is shown in Figure 4.2(b); the electrons are excited from region 1 to 2, across a width of $\sim k_B T$.

4.1.4 Heat Capacity of the Free Electron Gas

If the electron gas consists of classical particles, the kinetic energy for N electrons is expected to be $\frac{3}{2}Nk_B T$, and hence the heat capacity to be $\frac{3}{2}Nk_B$. However, experimentally, the heat capacity due to the free electrons in metals is found to be less than 1% of this value.

- The reason is that the Pauli exclusion and the Fermi–Dirac distribution allow only those electrons within an energy range $k_B T$ of the Fermi level to be excited thermally. The fraction excited is $\frac{k_B T}{\varepsilon_F} = \frac{T}{T_F}$ (see Figure 4.2(b)).
- Therefore, the electronic heat capacity, $C_{el} \sim Nk_B \frac{T}{T_F}$. A formal calculation gives, $C_{el} = \frac{\pi^2}{2}Nk_B\frac{T}{T_F}$. The heat capacity due to free electrons is proportional to temperature.
- The heat capacity of metals has contributions from the lattice as well as the free electrons so that, $C = C_{el} + C_{lattice} = aT + bT^3$.

4.1.5 Electrical Conduction and Ohm's Law

Electrical conductivity is one of the most basic of all the properties of materials. The phenomenon can be visualized using a simple approach in terms of the free electron theory.

- Momentum of free electrons is obtained as: $-i\hbar \nabla \psi_k (r)$ $= -i\hbar \nabla e^{ik \cdot r} = \hbar k \psi_k (r)$.
- The momentum, $\hbar k = mv$.
- In an electric field, E, the force acting on an electron is $-eE$ (Lorentz force with no magnetic field). Therefore, from Newton's second law of motion,
  - $m\dfrac{\partial v}{\partial t} = \hbar\dfrac{\partial k}{\partial t} = -eE.$

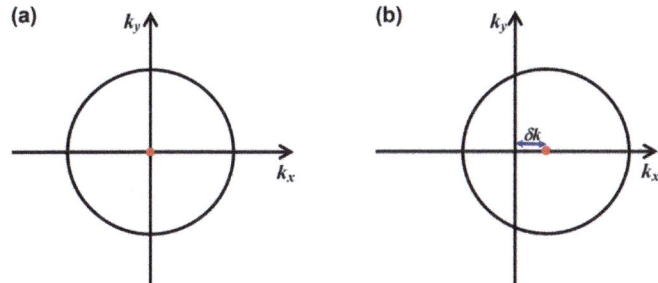

Figure 4.3   Fermi sphere (a) at time, $t=0$ and (b) under the influence of a force (F) due to an applied electric field, at time t; $\delta k = \dfrac{Ft}{\hbar}$.

- ○ Integration of the second equation from time $=0$ to time $=t$ gives, $k(t) - k(0) = \delta k = -\dfrac{eEt}{\hbar}$. This can be visualized as the displacement of the Fermi sphere in k-space at a uniform rate by the applied electric field (Figure 4.3).
- ○ At $t=0$, for every electron at wave vector k, there is one at wave vector, $-k$, so that there is no net momentum. When E is applied (force, $F=-eE$), all the momenta k are modified to $k+\delta k$. The net momentum for N electrons is $N\hbar\delta k$ and the increase in energy, $N\dfrac{\hbar^2(\delta k)^2}{2m}$.
- Acceleration of the electrons under an applied electric field does not continue forever, as the electrons are scattered by (collide with) lattice vibrations (phonons), impurities and defects, in an average time, τ (collision time). In other words, the Fermi sphere displacement reaches a steady state in time, τ.
  - ○ The change in wave vector, $\delta k = -\dfrac{eE\tau}{\hbar}$.
  - ○ Similarly, integration from time, 0 to τ, and velocity, 0 to v (the steady state velocity reached due to collisional relaxation) gives, $v = -\dfrac{eE\tau}{m}$. This is known as the **drift velocity.**

- Electric current density (current per unit cross-sectional area) due to n particles (per unit volume) with charge q and velocity, v is given by, $j=nqv$.
  - ○ Therefore, the current due to n electrons with charge, $-e$, is:

  $$j=n(-e)\left(-\dfrac{eE\tau}{m}\right) \Rightarrow j = \dfrac{ne^2\tau}{m}E \text{ (Ohm's law)}$$

Note: The usual formulation of Ohm's law (using voltage, V, current, I, and resistance, R) can be shown as follows:

$$j = \sigma E = \frac{E}{\rho} \qquad \Rightarrow \frac{I}{A} = \frac{V}{\rho l} \qquad \Rightarrow V = I\frac{\rho l}{A} = IR$$

- As $j = \sigma E$, the **conductivity**, $\sigma = \dfrac{ne^2\tau}{m} = ne\mu$, where $\mu = \dfrac{e\tau}{m}$ is known as the **mobility**. The higher the charge, the higher is the force exerted by the external field and hence the larger the mobility. Higher τ implies longer interaction of the field with the particle and hence higher mobility. For a fixed force, acceleration is inversely proportional to the mass; hence a heavier particle would possess lower mobility.
- Resistivity can be defined similarly, $\rho = \dfrac{1}{\sigma} = \dfrac{m}{ne^2\tau}$.
- $\sigma \sim 10^6\text{--}10^7\ \Omega^{-1}\,\text{cm}^{-1}$ for metals, whereas $\sigma \sim 10^{-12}\text{--}10^{-14}\ \Omega^{-1}\,\text{cm}^{-1}$ for insulators, at ambient (room) temperature.
- The resistivity arises due to lattice vibrations (L) and impurities/ defects (i). Hence the resistivity can be expressed as $\rho = \rho_L + \rho_i$. This is known as **Matthiessen's rule**.
- ρ extrapolated to 0 K gives ρ_i as ρ_L becomes zero at that temperature. $\rho(T)/\rho(0)$ provides an estimate of the purity of the sample.

4.1.6 Hall Effect

The Hall field is the electric field formed across two faces of a conductor, in the direction $j \times B$, when a current, j flows across a magnetic field, B (the direction is based on electrons being the charge carriers). In the example shown in Figure 4.4, the current flows along the x-axis, the magnetic field is applied along the z-axis and the electric field developed is along the y-axis; due to the negative charge of electrons, the field is directed along the negative y-direction.

- The Lorentz force on the electron in the presence of an electric field, E and magnetic field, B is given by: $F = -e\left(E + \dfrac{v \times B}{c}\right)$.

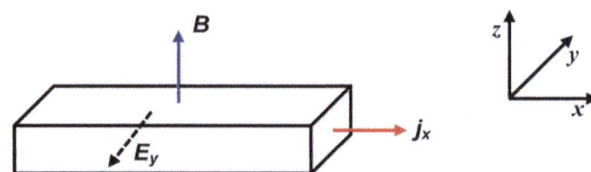

Figure 4.4 Standard geometry for the Hall effect description.

- The field developed, E_y can be determined from the expression for the velocity of electrons under the electric field, E_x and magnetic field B (applied along the z-direction). Following the expression for the velocity in Section 4.1.5, now with the effect of magnetic field added on, $\boldsymbol{v} = -\dfrac{e\tau}{m}\left(\boldsymbol{E} + \dfrac{\boldsymbol{v} \times \boldsymbol{B}}{c}\right)$.

- In the steady state when the electrons travel only in the x-direction, the velocity components, are:

$$v_x = -\frac{e\tau}{m}E_x$$

$$v_y = 0 = -\frac{e\tau}{m}\left(E_y - \frac{Bv_x}{c}\right)$$

(The above result follows from the determinant:

$$|\boldsymbol{v} \times \boldsymbol{B}| = \begin{vmatrix} x & y & z \\ v_x & 0 & 0 \\ 0 & 0 & B \end{vmatrix};$$

the x-component of the cross product $= 0$; y-component $= -Bv_x$.)

- Solving the above two equations:

$E_y = \dfrac{Bv_x}{c} = -\dfrac{eB\tau}{mc}E_x$. This is the **Hall field**.

- The Hall coefficient is defined as the electric field developed along the y-direction per unit current density in the x-direction and unit magnetic field in the z-direction:

$$R_{\mathrm{H}} = \frac{E_y}{j_x B} = -\frac{eB\tau E_x/mc}{(ne^2\tau E_x/m)\cdot B} \Rightarrow R_H = -\frac{1}{nec}$$

  ○ Clearly, the Hall coefficient is negative when electrons are the charge carriers.

  ○ Observation of a positive Hall coefficient in some metals (Al, In, *etc.*, under high magnetic fields) points to the presence of positive charge carriers. A theory beyond the free electron model is required to understand the phenomenon of holes.

4.1.7 Thermal Conductivity of Metals

As shown in Section 3.4 for phonons, the thermal conductivity due to free electrons can be written as: $\kappa_{\mathrm{el}} = \dfrac{1}{3}C_{\mathrm{el}}v_{\mathrm{F}}l$, where C_{el} is the heat capacity of free electrons, v_{F} is the Fermi velocity and l is the mean free path of electrons.

- Using the equation for heat capacity (Section 4.1.4) and replacing the Fermi temperature by Fermi energy and then Fermi velocity $(k_B T_F = \varepsilon_F = \frac{1}{2} m v_F^2)$,

$$C_{el} = \frac{\pi^2}{2} n k_B \frac{T}{T_F} = \frac{\pi^2}{2} \frac{n k_B^2 T}{\varepsilon_F} = \frac{\pi^2 n k_B^2 T}{m v_F^2}.$$

- Therefore, $\kappa_{el} = \frac{\pi^2}{3} \frac{n k_B^2 T}{m v_F^2} v_F l = \frac{\pi^2 n k_B^2 T \tau}{3m}$ ($\because$ mean free path, $l = \tau v_F$).

In pure metals, the electronic contribution to thermal conduction dominates over the phonon contribution. Thanks to the contribution of the free electrons, metals generally show much higher thermal conductivity than electrical insulators.

- **Wiedemann–Franz law:** The ratio of thermal conductivity to electrical conductivity of metals (at not too low temperatures) is proportional to the temperature, and the proportionality constant is a universal constant.
  - $\frac{\kappa_{el}}{\sigma} = \frac{\pi^2 n k_B^2 T \tau / 3m}{n e^2 \tau / m} = \frac{\pi^2}{3} \left(\frac{k_B}{e} \right)^2 T = LT$ ($L =$ Lorenz number $= 2.72 \times 10^{-13}$ esu deg^{-2}).
  - This law fails at low temperatures, since the collision time, τ becomes different for the thermal and electrical conduction processes.

4.2 Energy Bands

The state of an electron in a free atom is described by the quantum numbers, principal (n), orbital (l, azimuthal), magnetic (m_l) and spin (m_s). The energy of the free atom is determined by n and l; the $2l + 1$-fold degeneracy is lifted in the presence of a strong external field.

The orbitals of a free atom with different energies are designated as 1s, 2s, 2p, 3s, 3p, 3d, *etc.* A collection of N atoms which are far apart from each other would show N-fold degeneracy for each level (beyond the $2l + 1$), as the atoms are non-interacting. When atoms come close to each other (for example, as in a crystal), each one is subjected to the field due to its neighbors, and the N-fold degeneracy gets lifted depending on the extent of overlap and interaction of the corresponding orbitals.

A simple picture that can be visualized is that of a chain of H atoms and the band formed by the 1s orbitals. The most bonding molecular orbital (MO) with the periodicity of the lattice constant has the lowest energy; the most anti-bonding MO with the periodicity, twice that of the lattice constant, has the highest energy. The energy range between these two extremes is spanned by all the other MOs; the center of the band corresponds to the 1s orbital energy.

4.2.1 Energy Spectrum of Electrons in a Crystal

The Schrödinger equation for free electrons is given in Section 4.1.1. For electrons in a crystal it can be written as: $\left(-\dfrac{\hbar^2}{2m}\nabla^2 + U\right)\psi_k = \varepsilon_k\psi_k$, where $U = U_a + \delta U$ where U_a is the potential energy operator for the free atom and δU is the correction for the effect of the neighboring atoms in the crystal.

- If δU is neglected in a crystal, every level that was non-degenerate in the atom would have N-fold degeneracy (this is called transpositional degeneracy). When δU is turned on, this degeneracy is split and a band is formed; note that the number of levels in a band is finite.
- This is schematically illustrated for a sodium crystal in Figure 4.5.

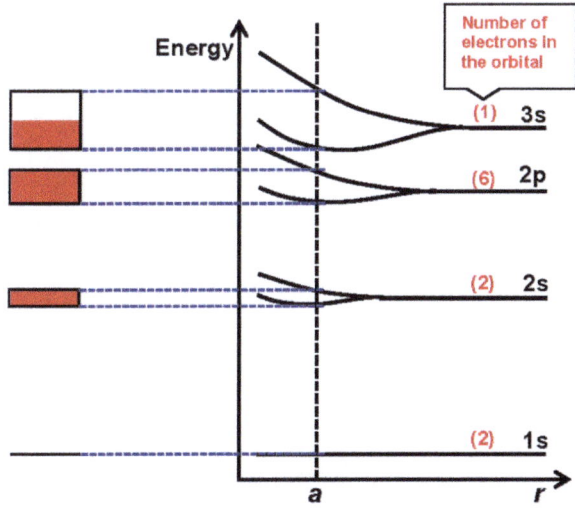

Figure 4.5 Schematic visualization of electronic band formation: splitting of the degenerate energy levels of the N-atom collection as the distance between the atoms (*r*) is reduced to the value, *a*, found in the crystal lattice. The number of electrons in the atomic energy levels of sodium (Na) are indicated, and the corresponding band occupation is shown by the red filling in the boxes on the left.

- The effect of the lattice field increases with increasing orbital size; broadening of the 3s orbital into a band is more pronounced than that of 2s, *etc.*
- Based on the electron filling, the 3s-band is half-filled. This gives rise to the metallic nature of Na; as a continuum of empty energy levels occur close above the top-most filled levels, the electrons can be accelerated under an external field without any energy (activation) barrier.
- The energy gaps between the bands indicate forbidden levels.
- If we extend a similar qualitative picture to Ca, all bands up to and including the 3s band are filled. However, the higher lying 3p band is sufficiently wide, so as to overlap with the 3s band, leading to an effectively partially filled band; this explains the metallic nature of Ca. The overlapping band systems are called '**semimetals**'.

If the band with the highest energy electrons (valence band) is fully filled, an energy gap separates it from the next empty band (conduction band); this leads to semiconducting or insulating behavior (depending on the energy gap). The qualitative band diagram for metals, semi-metals and semiconductors/insulators is shown in Figure 4.6.

A familiar chemical example that can be used to visualize the formation of electronic bands is provided by the extension of the π-electron system in ethylene to allyl radical to butadiene, *etc.*, to finally, poly(acetylene) which can be treated as a model 1-D crystal.

- Within a simple Hückel approach, the π–π* molecular orbital energy separation in ethylene is 2β, where β is the resonance integral.
- For each carbon atom (with a π-orbital) added, we introduce an additional energy level (molecular orbital), eventually ending up with an extended band having a width of 4β in poly(acetylene) (Figure 4.7).

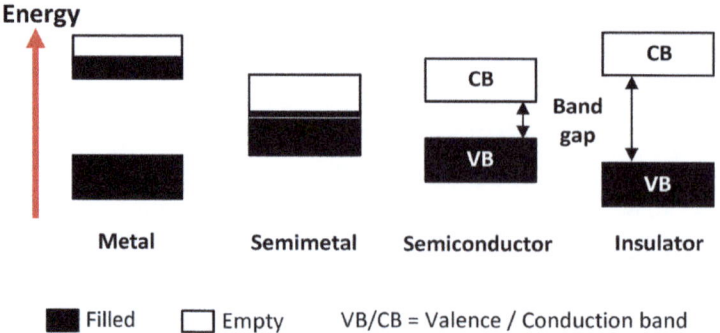

Figure 4.6   Qualitative picture of the band filling in metal, semimetal, semi-conductor (at 0 K) and insulator.

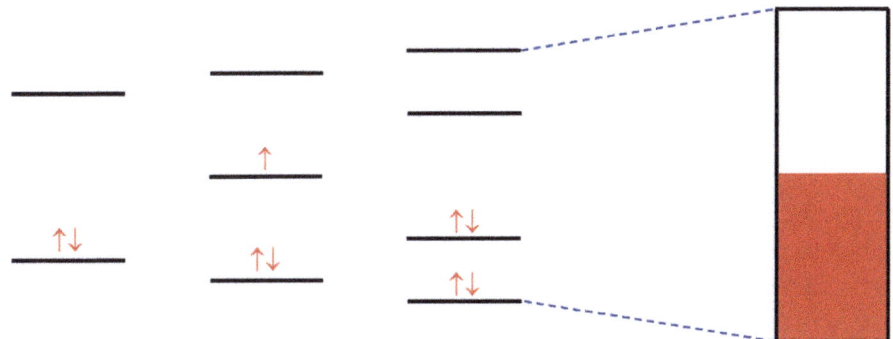

Figure 4.7 A qualitative view of band formation starting from molecular orbital energy levels.

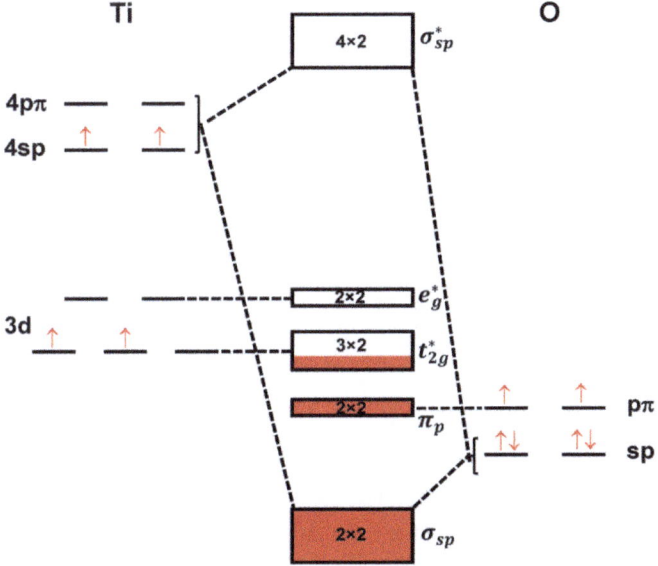

Figure 4.8 Qualitative band diagram of TiO; the numbers indicate the electron capacity (per formula unit) per band. Electron filling in the atomic orbitals (vertical arrows) and bands (red filling) are shown; the t_{2g}^* band is 1/3 filled.

- The electron filling suggests that the poly(acetylene) band is half-filled, resulting in a metal. We note later that the structure is prone to bond alternation that causes doubling of the unit cell length and a band gap opening up at the Fermi level, leading to the formation of a semiconductor.

Another example of a qualitative band diagram is shown for the case of TiO in Figure 4.8; partial filling of the t_{2g}^* band makes TiO a metal.

4.2.2 Electron Energy *Versus* Wave Wector: Origin of the Band Gap

As shown in Section 4.1.1, the free electron model (in 1-D) gives the wave function and energy as $\psi = c \cdot e^{ik.x}$ and $\varepsilon = \dfrac{\hbar^2 k^2}{2m}$ respectively. The energy dispersion is quadratic (parabola) and the probability of finding the electron, a constant everywhere, since, $\psi^*\psi = \int c \cdot e^{ik.x} \cdot c \cdot e^{-ik.x} = c^2$ (Figure 4.9).

The presence of a periodic potential due to the underlying lattice of positively charged nuclei leads to a significant change in the wave function as well as energy dispersion of the electrons; this can be understood in terms of the scattering of electron waves by the lattice, when the wave vector satisfies the Bragg (von Laue) condition.

- The von Laue condition for constructive interference of X-rays (Section 1.4.6) can be reformulated as follows.

$$\vec{k} - \vec{k'} = K \Rightarrow \vec{k} - \vec{K} = \vec{k'}$$

$$\left(\vec{k} - \vec{K}\right) \cdot \left(\vec{k} - \vec{K}\right) = \vec{k'} \cdot \vec{k'} \Rightarrow |\vec{k}|^2 - 2\vec{k} \cdot \vec{K} + |\vec{K}|^2 = |\vec{k'}|^2.$$

  Since the collision is elastic, energy and momenta are conserved. Therefore, $\left|\vec{k}\right|^2 = |\vec{k'}|^2$. Hence, $2\vec{k} \cdot \vec{K} = |\vec{K}|^2 \Rightarrow \vec{k} \cdot \left(\dfrac{1}{2}\vec{K}\right) = \left|\dfrac{1}{2}\vec{K}\right|^2$.

- The last equation is satisfied by any electronic wave vector, $\vec{k}$ incident on the perpendicular bisector of a reciprocal lattice vectors, $\vec{K}$, and hence will undergo Bragg reflection, as shown graphically in Figure 4.10(a).
- The perpendicular bisectors of the reciprocal lattice vectors thus define the Brillouin zones; the case of a 2-D square lattice is

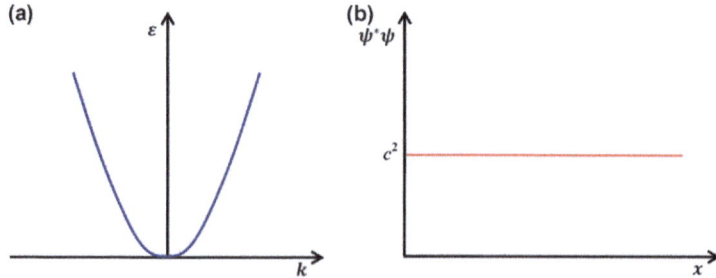

Figure 4.9 (a) Energy dispersion curve, and (b) spatial probability distribution for free electrons in a 1-D crystal; see text for the meaning of the symbols.

shown in Figure 4.10(b). Note that the Brillouin zones are Wigner–Seitz cells (Section 1.3.5) in reciprocal space.

The effect of the Bragg reflection at the Brillouin zone boundary, on the wave function can be visualized as follows, once again taking the 1-D example.

- Bragg reflection occurs for $k = \dfrac{\pi}{a}$; as $K = \dfrac{2\pi}{a}$, the reflected wave vector, $k' = -\dfrac{\pi}{a}$.
- At the boundary, the incident and reflected travelling waves ($e^{\frac{i\pi x}{a}}$ and $e^{-\frac{i\pi x}{a}}$) give rise to the standing waves, $\psi_{\pm} = e^{\frac{i\pi x}{a}} \pm e^{-\frac{i\pi x}{a}}$. These take the functional forms, $\cos\left(\dfrac{\pi x}{a}\right)$ and $\sin\left(\dfrac{\pi x}{a}\right)$ respectively.
- The probability of finding the electrons is $\cos^2\left(\dfrac{\pi x}{a}\right)$ and $\sin^2\left(\dfrac{\pi x}{a}\right)$ respectively. The former will show maxima at $x = na$ ($n = 0, 1, 2, \ldots$), the lattice points where the nuclei are located, whereas the latter has maxima at $x = \dfrac{(2n+1)a}{2}$, mid-way between the nuclei (Figure 4.11(a); compare with Figure 4.9(b)).
- ψ_{+} which piles up electron density at the positively charged nuclei is stabilized with respect to the free electron wave function which spreads out the electron density evenly. ψ_{-} is destabilized with respect to the free electron wave function. This leads to the energy gap opening up at the Brillouin zone boundary as shown in Figure 4.11(b) (compare with Figure 4.9(a)).

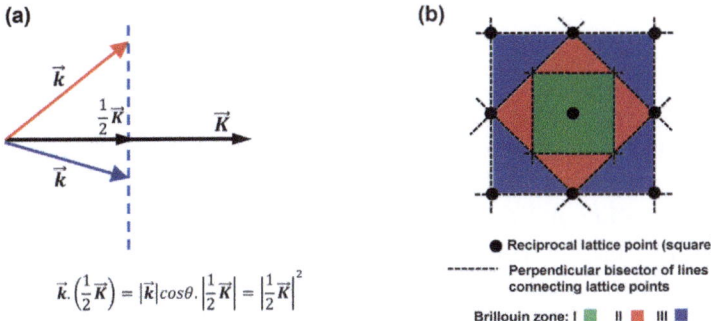

(a)

$$\vec{k}\cdot\left(\tfrac{1}{2}\vec{K}\right) = |\vec{k}|\cos\theta.\left|\tfrac{1}{2}\vec{K}\right| = \left|\tfrac{1}{2}\vec{K}\right|^2$$

(b)

- ● Reciprocal lattice point (square)
- - - - - Perpendicular bisector of lines connecting lattice points
- **Brillouin zone: I** ■ **II** ■ **III** ■

Figure 4.10 (a) Formulation of the von Laue condition showing that waves (electronic in the current context) with vectors incident on the perpendicular bisector of the reciprocal lattice vector undergo diffraction; (b) Brillouin zones defined as Wigner–Seitz cells in reciprocal lattice.

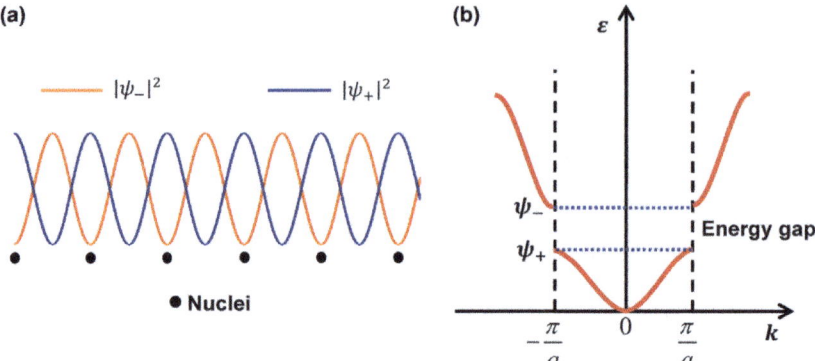

Figure 4.11　(a) Spatial probability distribution corresponding to the standing waves at the Brillouin zone boundary of a 1-D crystal. (b) Corresponding energy dispersion curve showing the energy gap opening up at the Brillouin zone boundary.

The Brillouin zones of a 3-D crystal are marked by boundaries in different directions where the energy gaps open up. The full band diagram would thus involve plots of ε *versus* different $\boldsymbol{k}$.

4.2.3　Bloch Functions

The general solution to the Schrödinger equation with a periodic potential (Section 4.2.1) is given by **Bloch functions**, $\psi_k(r) = u(r) \cdot e^{ik \cdot r}$, with the first part satisfying the condition, $u(r) = u(r + T)$ where T is the lattice translation.

- This shows that $\psi_k(r + T) = \psi_k(r) \cdot e^{ik \cdot T}$.
- For a 1-D system, for example: $\psi_k(x + na) = u(x + na) \cdot e^{ik \cdot (x + na)}$
  $= u(x) \cdot e^{ik \cdot x} e^{inka} = \psi_k(x) \cdot e^{inka}$ (note that $k = s\left(\dfrac{2\pi}{Na}\right)$ where $s = 0$, $\pm 1, \pm 2, \ldots$).

A physical picture of the Bloch function can be developed using the idea of molecular orbitals; the following illustration is based on the discussion of polyacetylene bands.[3] In principle, the polyacetylene structure can be considered as 'regular' or 'dimerized' (Figure 4.12(a)). In the 'dimerized' structure observed experimentally under ambient conditions, the bonds are long and short alternately.

- Molecular orbitals can be constructed using ethylene π and π^* MOs as the basis functions (Figure 4.12(b)). Four specific cases drawn assuming an underlying dimerized structure are shown in

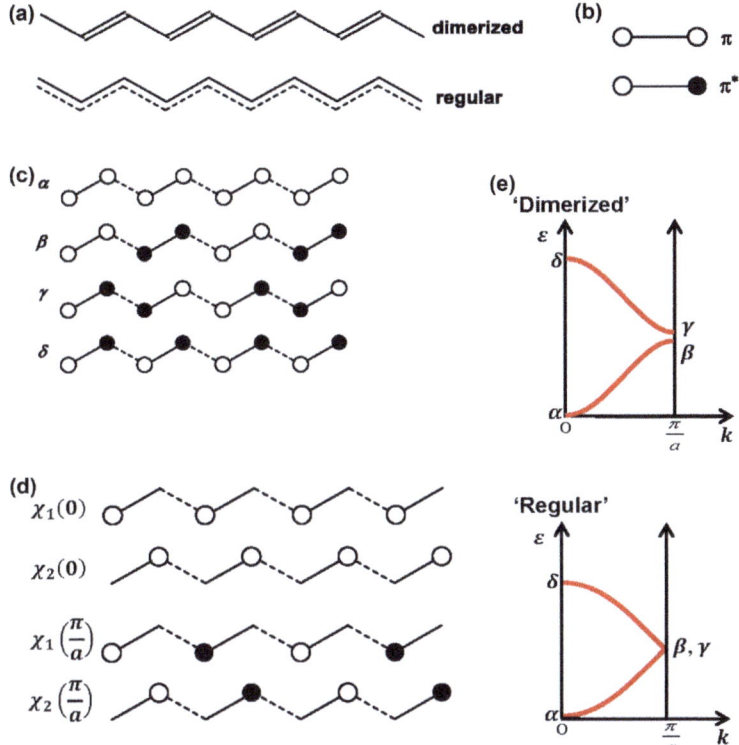

Figure 4.12 (a) Structure of polyacetylene. (b) Ethylene π MOs used as the basis. (c) Selected MOs of polyacetylene constructed using the basis in (b). (d) Bloch orbitals of polyacetylene created from the p_z-orbitals of the two C atoms in the unit cell. (e) Band diagram showing the placement of the MOs (in the two structures of polyacetylene). Drawn schematically based on Figures 1, 4 and 5 in ref. 3.

Figure 4.12(c); α and β are obtained by adding π MOs in phase and out of phase respectively, and γ and δ are generated similarly from the π* MOs. Based on the nodal pattern, the energy gradation would be, $\varepsilon_\alpha < \varepsilon_\beta < \varepsilon_\gamma < \varepsilon_\delta$. It can be noted that $\varepsilon_\beta < \varepsilon_\gamma$ because the bonding interaction (no node) coincides with the shorter bond in β, whereas it coincides with the longer bond in γ. If the structure was regular, these MOs would be degenerate.

- An alternate description of the MOs would be in terms of Bloch functions (Bloch orbitals) of the form, $\chi_\mu(k) = \sum_n \varphi(\mu)e^{ikna}$. φ are the atomic orbitals (p_z) at the sites $\mu = 1,2$ in the unit cell (ethylene unit), k is the wave vector of the Bloch orbital, n represents the count of the unit cell from an arbitrary origin, and a is the unit cell length; na thus represents the position of the nth unit cell. The Bloch functions for the wave vectors, $k = 0$ and $k = \dfrac{\pi}{a}$ (Brillouin zone boundary) are shown in Figure 4.12(d);

$\chi_\mu(0) = \displaystyle\sum_n \varphi(\mu)$, with the atomic orbitals added in phase, and

$\chi_\mu\left(\dfrac{\pi}{a}\right) = \displaystyle\sum_n \varphi(\mu)e^{in\pi}$, with the atomic orbitals added with phase changing by π alternately.

- Close inspection of the MOs in Figure 4.12(c) and the Bloch orbitals in Figure 4.12(d) reveals the following relations:

$$\alpha = \chi_1(0) + \chi_2(0); \qquad \beta = \chi_1\left(\dfrac{\pi}{a}\right) + \chi_2\left(\dfrac{\pi}{a}\right)$$

$$\gamma = \chi_1\left(\dfrac{\pi}{a}\right) - \chi_2\left(\dfrac{\pi}{a}\right); \qquad \delta = \chi_1(0) - \chi_2(0).$$

- Considering the energy gradation, and the wave vectors, these MOs can be positioned on a band diagram as shown in Figure 4.12(e); this is equivalent to Figure 4.11(b), with the upper band folded back, as $k = \dfrac{2\pi}{a}$ is equivalent to $k = 0$.

- It may be noted that if the structure was 'regular', the unit cell length would be $a/2$ and the Brillouin zone boundary occurs at $\dfrac{2\pi}{a}$; hence no gap would occur at $k = \dfrac{\pi}{a}$.

- The 'regular' structure with half-filled band will be a metal; the Peierls instability leads to dimerization (electronic energy gain overwhelms the cost of lattice distortion at the temperature of interest) and causes the band splitting. The 'dimerized' structure makes polyacetylene a semiconductor.

4.2.4 Effective Mass

As shown in Sections 4.1.1 and 4.2.2, the energy *versus* wave vector relation for a free electron is given by $\varepsilon(k) = \dfrac{\hbar^2 k^2}{2m}$. Since, $\dfrac{\partial^2 \varepsilon}{\partial k^2} = \dfrac{\hbar^2}{m}$, the mass can be defined as $m = \hbar^2 \bigg/ \dfrac{\partial^2 \varepsilon}{\partial k^2}$. However, the energy dispersion curve for an electron in a periodic potential is no longer described by the simple parabola, as shown in Section 4.2.2.

- Modification of the energy dispersion function, especially near the zone boundary, is taken into account by defining an **effective mass** as: $m^* = \hbar^2 \bigg/ \dfrac{\partial^2 \varepsilon}{\partial k^2}$.

- The electron in a periodic potential is described in terms of a free electron picture, but with varying effective mass. A band with low curvature is associated with a large effective mass and *vice versa*.
- The effective mass becomes negative at the band maximum.

The bizarre concept of a variable, and even negative mass has the following physical basis.

- For free electrons, application of an external force (by an electric field), just increases their kinetic energy.
- The acceleration imparted to an electron in a periodic lattice by an external field could be greater or lesser than that imparted to a free electron, depending on the actual wave vector of the electron. In the former case, the electron appears to be lighter than a free electron, and in the latter case, heavier.
- For electrons near the bottom of the lower band, the wave function is free electron like, and the momentum is $\hbar k$.
- At the top of the band, near the Brillouin zone boundary, an increase in momentum from k to $k + \Delta k$ by an applied field takes the electron wave closer to Bragg reflection, and thus to a decrease in the overall momentum (k and $-k$ get superposed, leading to standing waves). As $\hbar \dfrac{\partial k}{\partial t} = m^* \dfrac{\partial v}{\partial t}$ is positive, and $\dfrac{\partial v}{\partial t}$ is negative, the effective mass m^* is negative.

4.3 Metals and Semiconductors

Metals and semiconductors form some of the most prominent materials of technological relevance. Insight into their fundamental characteristics is necessary to develop an understanding of various related classes of materials. Figure 4.6 shows the difference in the band filling in metals and semiconductors.

- Due to the partial filling of the band in metals, the electrons at the Fermi energy have directly accessible energy levels, and hence can be accelerated in an electric field without having to overcome any activation barrier; the electrical transport is unactivated.
- In semiconductors, the electrons in the valence band have to be promoted to the conduction band, for electrical conduction to take place; this leads to activated transport.

- In both cases, the electron mobility is affected by lattice vibrations and hence decreases with increasing temperature. The number of charge carriers is largely independent of temperature in metals, whereas it increases with temperature in semiconductors. These factors lead to contrasting dependence of electrical conductivity on temperature in metals and semiconductors.

4.3.1 Metals

As discussed in Section 4.1.5, electrical conductivity is given by $\sigma = ne\mu$.

- As n is not temperature dependent, the electrical conductivity varies with temperature as the mobility does. The temperature dependence of the mobility of free electrons in metals is shown in Figure 4.13(a); the consequent variation of electrical resistivity with temperature is shown in Figure 4.13(b).
- At temperatures above the Debye temperature, the carrier mobility varies with temperature as T^{-1}, and below the Debye temperature, as T^{-5}.
- At very low temperatures where the phonons die down fully (phonon numbers vanish even for the lowest energy acoustic modes), the carrier scattering is only due to lattice impurities and defects, and the mobility becomes constant; the corresponding resistivity is ρ_i (see Section 4.1.5).

The experimental plot of the resistivity of gold *versus* temperature in Figure 4.14 shows the linear regime. The resistivity of good metals at ambient temperatures is $\sim 10^{-6}\ \Omega\,\text{cm}$; values for some typical metals are collected in Table 4.1.

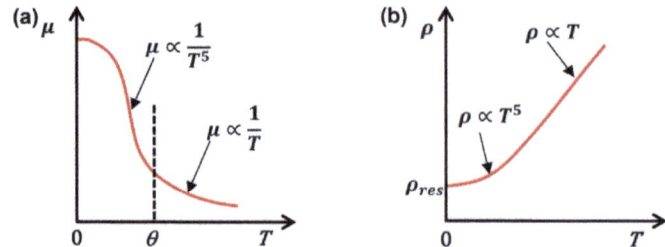

Figure 4.13 Temperature dependence of the (a) the mobility of free electrons, and (b) electrical resistivity of pure metals. Drawn schematically based on Figures 6.9 and 6.10 in ref. 4.

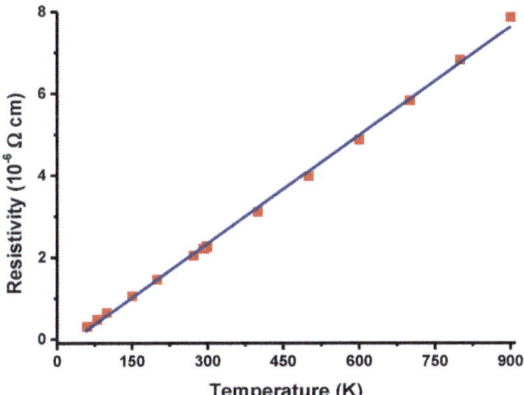

Figure 4.14 Temperature dependence of the resistivity of gold down to 60 K. Drawn based on the data in ref. 5; least square fit of the data to a straight line (blue) is also indicated.

Table 4.1 Resistivity of some metals.

Metal	ρ $(10^{-6}\,\Omega$ cm) at 20 °C
Silver	1.59
Copper	1.68
Gold	2.44

4.3.2 Intrinsic Semiconductors

Insulators and semiconductors have completely filled valence and empty conduction bands (at 0 K), separated by the band gap, E_g (Figure 4.6).

- Typically, insulators have an $E_g > 3$ eV and ambient temperature resistivity of $\sim10^{14}$–10^{22} Ω cm.
- Semiconductors on the other hand have $E_g \lesssim 1$ eV and ambient temperature resistivity of $\sim10^{-2}$–10^{6} Ω cm.
- Division of insulators and semiconductors is rather qualitative. Due to the strong temperature dependence, the resistivity can vary significantly; for example, diamond with an $E_g \sim 5$ eV has a resistivity of 10^{10} Ω cm under ambient conditions, but has a resistivity of ~50 Ω cm at 600 K.

Intrinsic semiconductors are based on pure materials; those with dopants of foreign atoms are called **extrinsic semiconductors**. Examples of intrinsic semiconductors include pure elements like Si and Ge, and compounds like GaAs and InSb. Their band gaps and ambient temperature resistivity are shown in Table 4.2.

Band gap is the difference in energy between the lowest point in the conduction band (**conduction band edge**) and the highest point in the valence band (**valence band edge**) in k-space.

- Depending on the specific band structure of the material, these may occur at the same point in k-space (Figure 4.15(a)), or be

Table 4.2   Band gap and typical ambient temperature resistivity of some semiconductors.

Material	E_g (eV)	ρ (Ω cm)
Si	1.12	$\sim 10^5$
Ge	0.66	~ 50
GaAs	1.43	$\sim 10^{-3}$–10^8
InSb	0.17	$\sim 10^{-3}$

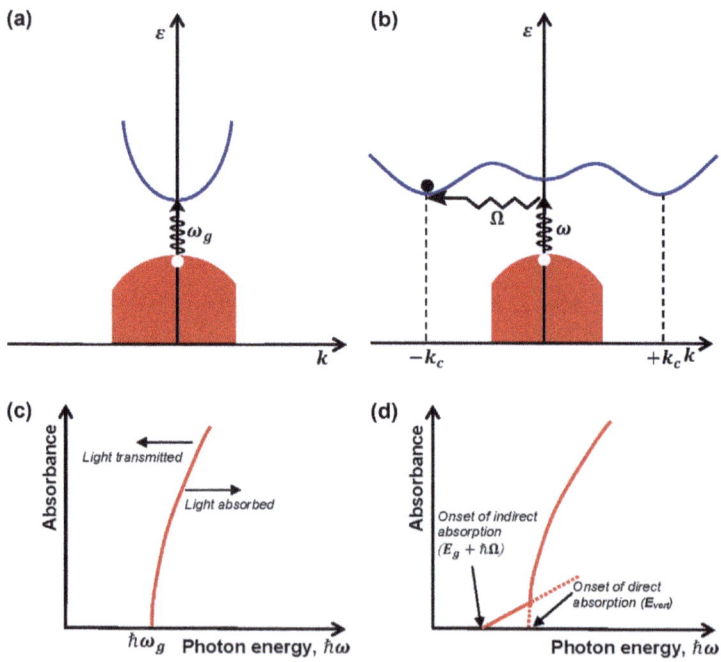

Figure 4.15   Filled valence band (red) and empty conduction band (blue) edges in (a) direct and (b) indirect band gap semiconductors (black and white filled circles symbolize the excited electron and its absence respectively); the photon frequencies (ω_g, ω) are shown in (a) and (b) respectively, and the phonon frequency (Ω) and wave vector ($-k_c$) in the latter. Optical absorption of (c) direct and (d) indirect band gap semiconductors; the onset of indirect and direct absorptions are shown in the latter.

separated by a wave vector, k_c (Figure 4.15(b)). The two cases are called **direct and indirect band gap semiconductors** respectively.

- The nature of the optical absorption in the two cases are distinct as shown in Figure 4.15(c) and (d) respectively. In the direct absorption process, steady absorption starts at the threshold frequency, ω_g given by $E_g = \hbar\omega_g$. In the indirect absorption, there is an onset of absorption at frequency, ω given by $E_g + \hbar\Omega = \hbar\omega$, where Ω corresponds to the frequency of a phonon that is emitted together with the electron excitation from the valence band to the conduction band. Momentum conservation shows that $K + k_c = k_{photon}$ where the wave vector of the phonon is K (see Figure 4.15(b)).
- At high temperatures, when a higher population of phonons is present (vibrational excitation), the threshold energy could be $E_g - \hbar\Omega = \hbar\omega$; a photon + phonon are absorbed to create the electron excitation.

4.3.3 Holes

The concept of holes is central to an understanding of semiconductors. Vacant orbitals in a band, resulting from excitation or removal of electrons, are known as holes. The relevant characteristics of **holes in relation to the electron before it was removed**, are listed below and summarized in Table 4.3.

- The total wave vector of electrons in a filled band is zero, $\Sigma k = 0$. Therefore, if an electron is missing from an orbital with wave vector, k_e, the total wave vector of the system is now $-k_e$, which is attributed to the hole. In other words, the wave vector of the hole, $k_h = -k_e$. This is shown graphically in Figure 4.16(a).
- Conventionally, the energy of the top of the filled band is taken to be zero. Therefore, the lower in the band that the electron is missing, the higher the energy of the resulting system, the hole. The energy of the hole is opposite in sign to the energy of the missing electron, as more work has to be done to remove an electron

Table 4.3 Relations between the properties of a hole and the missing electron.

Property	Relation
Wave vector	$k_h = -k_e$
Energy	$\varepsilon_h(k_h) = -\varepsilon_e(k_e)$
Velocity	$v_h(k_h) = v_e(k_e)$
Mass	$m_h = -m_e$

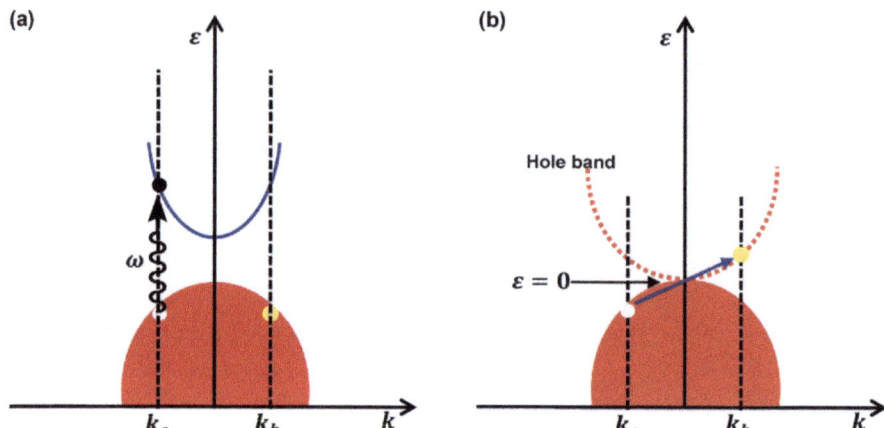

Figure 4.16 (a) Excitation of an electron from the valence band with wave vector k_e (white filled circle) by absorption of photon, $\hbar\omega$ gives an electron in the conduction band with the same wave vector (black filled circle); the wave vector of the valence band, $-k_e$, is assigned to the hole, which can be represented by the wave vector, k_h (yellow filled circle). (b) The hole formation can be described schematically using a hole band constructed based on the relations, $k_h=-k_e$ and $\varepsilon_h(k_h)=-\varepsilon_e(k_e)$; the hole band is an inverted (about both axes) image of the electron band.

from a lower energy state than from a higher one. This means that $\varepsilon_e(k_e)=\varepsilon_e(-k_e)=-\varepsilon_h(-k_e)=-\varepsilon_h(k_h)$; note that the states corresponding to the k_e and $-k_e$ vectors are degenerate. The relations between the wave vector and energy of the missing electron and the resulting hole can be shown schematically by drawing a hole band, which is a total inversion of the corresponding electron band (Figure 4.16(b)).

- Since $\varepsilon(k)=\dfrac{\hbar^2 k^2}{2m}$, the group velocity is given by $v(k)=\dfrac{\hbar k}{m}=\dfrac{1}{\hbar}\nabla\varepsilon(k)$, slope of the band. Figure 4.16(b) shows that, $\nabla\varepsilon_h(k_h)=\nabla\varepsilon_e(k_e)$; hence $v_h(k_h)=v_e(k_e)$. In Figure 4.16(b), both correspond to positive slopes. It may be noted that this relates to the electron before its removal or excitation. The electron residing in the conduction band **after excitation** (orbital shown by the black circle in Figure 4.16(a)), will indeed have a velocity that has a sign opposite to that of the hole.

- From Figure 4.16(b), it can also be inferred that the effective mass (given by the curvature of the band as shown in Section 4.2.4) of the missing electron (before removal) and that of the hole formed are opposite in sign $m_h=-m_e$.

4.3.4 Electrical Conductivity of Intrinsic Semiconductors

Under an electric field, the current due to the electron excited to the conduction band and the hole formed in the valence band add to each other (Figure 4.17).

- Assume that the field is in the x-direction (*i.e.*, only E_x present). The electron due to its negative charge, moves opposite to the direction of the field (velocity in the $-x$ direction), and hence contributes to the current in the $+x$ direction ($j = nqv$).
- The hole has effective charge $+e$. This can be shown from the equation of motion: $\hbar \dfrac{\partial k_e}{\partial t} = -eE = \hbar \dfrac{\partial(-k_h)}{\partial t} \Rightarrow \hbar \dfrac{\partial k_h}{\partial t} = eE$. As the velocity of the hole is opposite to that of the electron, the current due to the hole is in the $+x$ direction, as the charge of the hole is also opposite to that of the electron.

The number of conduction electrons (n_i) and holes (p_i) in an intrinsic semiconductor are equal and depends on the absolute temperature (T) as well as the band gap (E_g) as:

$$n_i = p_i = 2 \left(\frac{2\pi \sqrt{m_n^* m_p^*} k_B T}{h^2} \right)^{3/2} e^{-E_g/2k_B T}$$

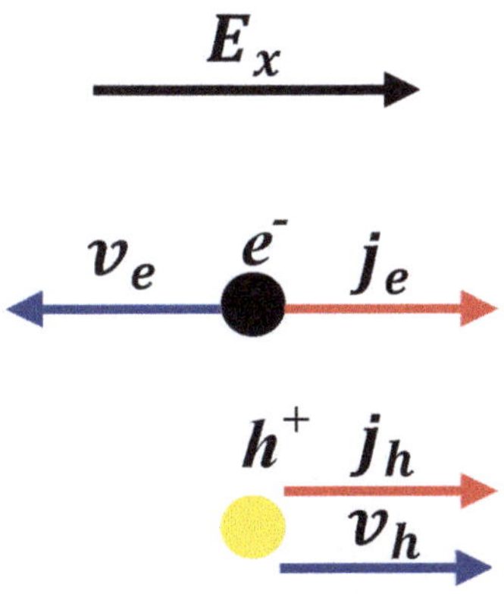

Figure 4.17 Motion (represented by velocity, **v**) of electron (e⁻, in the conduction band) and the hole (h⁺) created by its excitation (in the valence band), and their contributions to the current (*j*) in an external electric field, E_x. Holes move in the opposite direction as their effective mass is negative; they can be viewed as electrons with negative charge and negative mass or positive charge and positive mass.

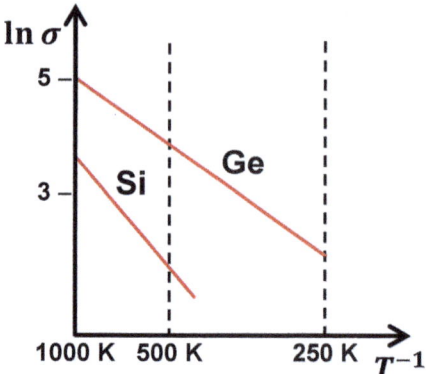

Figure 4.18 Schematic plot of the temperature (T) dependence of the electrical conductivity (σ) of semiconductors.

where m_n^* and m_p^* are the effective mass of electrons and holes. The Fermi level is at $\dfrac{E_g}{2}$ at 0 K; it varies with temperature depending on the actual band shape and hence with the relative values of m_n^* and m_p^*. At relatively high temperatures, the mobility of electrons and holes in a semiconductor depends on the temperature as: $\mu \propto T^{-3/2}$. The conductivity of pure and perfect crystals is given by $\sigma_i = \sigma_n + \sigma_p = n_i e (\mu_n + \mu_p)$.

- Combining the expressions for the number of charge carriers and the mobility, one gets the familiar expression for the conductivity of semiconductors:

$$\sigma_i = \sigma_0 e^{-E_g/2k_B T}.$$

- This is also expressed commonly as $\ln \sigma_i = \ln \sigma_0 - \dfrac{E_g}{2k_B T}$. Plot of $\ln \sigma_i$ *versus* $\dfrac{1}{T}$ gives a straight line with the slope equal to $\dfrac{E_g}{2k_B}$ (Figure 4.18).

4.3.5 Extrinsic Semiconductors

If an instrinsic semiconductor is doped with impurity atoms which can donate electrons into the conduction band or accept electrons from the valence band (creating holes) even under ambient temperature conditions, the mobile charge carrier concentration increases significantly. These are called extrinsic semiconductors.

Consider a Ge lattice (diamond structure) doped with a few percent of As atoms which occupy the lattice sites. The four valence electrons

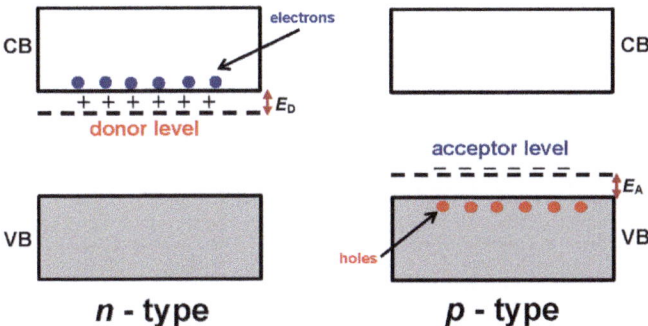

Figure 4.19 Schematic band diagrams of n- and p-type semiconductors; VB = valence band, CB = conduction band.

of Ge are engaged in the tetrahedrally oriented bonds. Since As has five valence electrons, the extra electron is available as a charge carrier; ionization of As is assisted by the dielectric nature of Ge ($\varepsilon_r = 16$) which decreases the attractive interaction between the As core and the valence electron. The donation of electrons into Ge can be represented in a band diagram (Figure 4.19); the discrete donor level (and not a band) arises due to the very low concentration of the dopant As atoms; the energy difference with the conduction band edge is, $E_D \sim 0.01$ eV. As-doped Ge is an example of an n-type semiconductor.

The contrasting situation of hole formation in the valence band arises due to the acceptor level introduced by doping the trivalent In atoms (Figure 4.19); the energy difference with the valence band edge is, $E_A \sim 0.01$ eV. In-doped Ge is a p-type semiconductor.

At relatively low temperatures, charge carriers in the semiconductor arise primarily from the dopant atoms. When the concentration of dopants in the n-type and p-type semiconductors are N_D or N_A respectively, the number of charge carriers, electrons (n) or holes (p) at temperature T are given by the equations:

$$n = \sqrt{2N_D} \left(\frac{2\pi m_n k_B T}{h^2} \right)^{3/2} e^{-E_D/2k_B T}$$

$$p = \sqrt{2N_A} \left(\frac{2\pi m_p k_B T}{h^2} \right)^{3/2} e^{-E_A/2k_B T}$$

The electrical conductivity is given by: $\sigma_{imp} = \sigma_{imp}^0 e^{-E_X/2k_B T}$, where $E_X = E_D$ or E_A. The pre-exponential temperature factor of n (or p) cancels the temperature dependence of mobility.

On increasing the temperature, a point is reached when all the donor/acceptor atoms are ionized (to form cations or anions

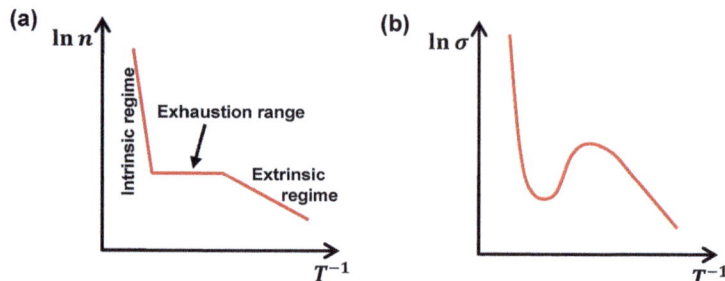

Figure 4.20   Schematic plots of the temperature (T) dependence of the (a) charge carrier concentration (n), and (b) electrical conductivity (σ) of extrinsic semiconductors.

respectively). This is known as the exhaustion range. At higher temperatures, electrons are excited from the valence band to the conduction band, and intrinsic conductivity sets in. This is shown schematically in Figure 4.20.

4.3.6   p–n Junctions

The majority carriers in n-type semiconductor are electrons, and charge neutrality is ensured by the equal number of positive charges on the donor sites. Similarly, the majority carriers in p-type semiconductor are holes, and charge neutrality is ensured by the equal number of negative charges on the acceptor sites. When p-type and n-type semiconductors are in contact, limited diffusion of the majority carriers across the junction occurs (Figure 4.21). The resulting excess negative and positive charges on the p and n regions creates an electrical double layer; the associated electric field directed from the n to the p region acts as a barrier for further carrier diffusion.

- A forward bias (positive and negative voltage applied to the p- and n-regions respectively) reduces the potential barrier (Figure 4.22(a)) and enhances the current flow, whereas, a reverse bias increases the potential barrier (Figure 4.22(b)) and reduces the current flow; the resulting I–V plot (Figure 4.22(c)) shows that the p–n junction acts as a **rectifier**.
- A **bipolar junction transistor (BJT)** with a p–n–p structure acts as a switch, lowering and raising the potential barrier between the *emitter* and the *collector* (the p-regions) allowing current to flow or stop, by the application of a suitable voltage at the *base* (n-region).

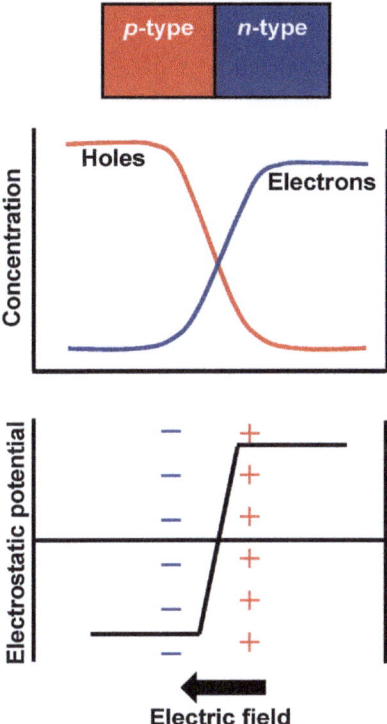

Figure 4.21 p–n junction, electron–hole concentration at the junction at equilibrium and the resulting electrical double layer, potential gradient and electric field.

4.3.7 Thermoelectric Effects

Thermoelectric effects arise due to a temperature gradient across a material causing the generation of an electric voltage and *vice versa*.

- **Seebeck effect** is the build-up of the voltage (V) due to the temperature difference (ΔT); the Seebeck coefficient (also known as thermopower), S is defined as: $S = \dfrac{V}{\Delta T}$. The voltage arises due to the difference in the rates of diffusion of charge carriers from the two ends maintained at different temperatures. The higher energy electrons (holes) move from the hotter end to the colder end, leading to a build-up of the voltage.
- A thermocouple works on the principle that the voltage V_{12} developed by placing two junctions of conductors, A and B at two different temperatures (T_1 and T_2) is proportional to the difference in the two temperatures.

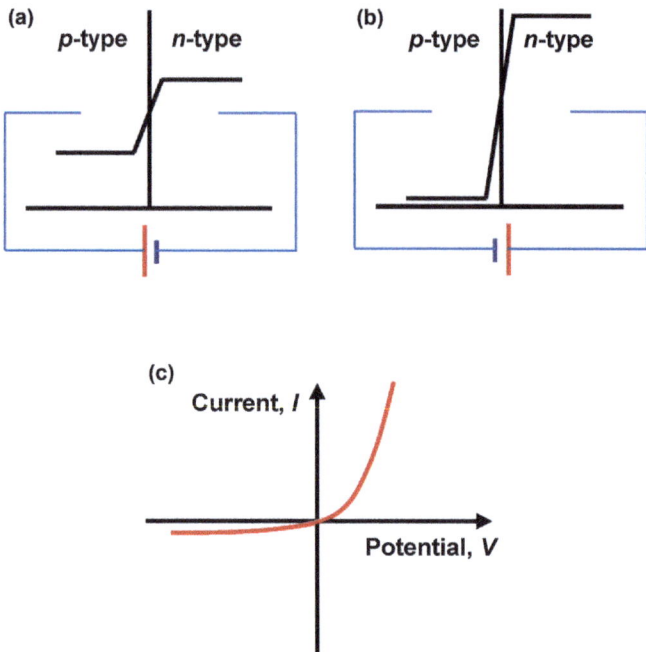

Figure 4.22 Change of the potential barrier at the p-n junction upon (a) forward, and (b) reverse bias. (c) The rectifying action of a p-n junction resulting from this.

$$V_{12} = \int_{T_1}^{T_2} [S_B(T) - S_A(T)]\mathrm{d}T.$$ Assuming that the Seebeck coefficient is nearly constant in the temperature range of interest, the equation reduces to:

$V_{12} = [S_B - S_A] \cdot \Delta T$. The schematic representation of a thermocouple (A–B) in Figure 4.23(a) shows that measurement of the voltage and knowledge of a reference temperature (T_1) allows determination of the sample temperature (T_2). Chromel–alumel is a well-known thermocouple used typically in the 200–1200 °C range; they are alloys made up primarily of $Ni + Cr$ and $Ni + Mn + Al + Si$ respectively.

- When a current flows through a junction of two conductors, A and B, heat may be generated or removed from the junction; this is known as the **Peltier effect**. The heat generated in unit time, $\dot{Q}$ is related to the current I and the Peltier coefficients (Π) of the conductors by the equation, $\dot{Q} = (\Pi_A - \Pi_B)I$.

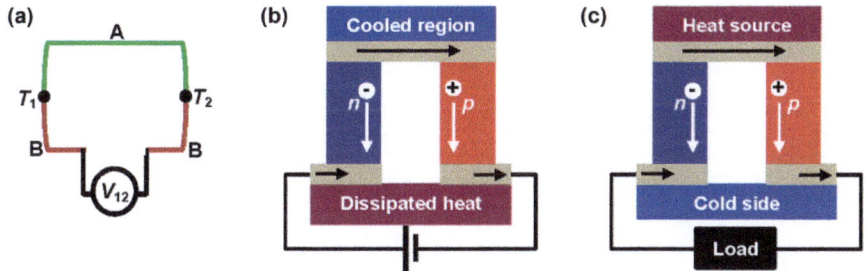

Figure 4.23 (a) Schematic diagram of a thermocouple, A–B with junctions maintained at temperatures, T_1 and T_2, developing a voltage, V_{12}. (b) Principle of Peltier cooling: application of the voltage as shown causes electrons in the n-type semiconductor (dominant carrier) to flow opposite to the current (black arrow; grey blocks are metallic interconnects), and holes in the p-type semiconductor (dominant carrier) in the direction of the current, both removing heat from the region being cooled. (c) Power generation resulting from the flow of electrons in the n-type semiconductor driven by the heat source towards the cold end: holes in the p-type semiconductor flow in the same direction as the current (black arrows).

- The Peltier effect is used in cooling devices (Peltier cooling) through an ingenious application of p–n junctions (Figure 4.23(b)). The current applied from a power source causes electron and hole flows that transport the heat away from the region to be cooled.
- A reverse process in which the dominant carriers flow from the hot end to the cold end, generates a current to run a load (Figure 4.23(c)); the thermoelectric effect is thus used to develop a power generator.

4.3.8 Hopping Semiconductors

Following the qualitative approach to band diagrams (Section 4.2.1), for example the case shown for TiO (Figure 4.8), the series of transition metal monoxides are expected to show metal/semiconductor behavior as shown in Table 4.4. However, the experimental observations contradict these predictions in most cases. The anomaly arises due to the relevant metal d-orbitals not forming extended bands.

- If the experimental cation–cation distance, R_E is larger than a critical value, R_C, the d-orbital overlaps are poor and the band picture becomes invalid due to strong electron–electron

Table 4.4 d-orbital filling, expected and observed characteristic (M = metal, SC = semiconductor), and cation–cation distances (R_E = experimental, R_C = calculated threshold value) of transition metal monoxides.

Metal oxide	TiO	VO	MnO	FeO	CoO	NiO
Filling of t_{2g}^*	2/6	3/6	5/6	6/6	6/6	6/6
e_g^*	—	—	—	—	1/4	2/4
Expected	M	M	M	SC	M	M
Observed	M	SC	SC	SC	SC	SC
R_E (Å)	2.89	2.89	3.14	3.03	3.01	2.95
R_C (Å)	3.02	2.92	2.66	2.95	2.87	2.77

interaction; an empirical relation developed by Goodenough[6] shows that $R_C = 3.2 - 0.05m - 0.03(Z - Z_{Ti}) - 0.04\{S_i \ (S_i + 1)\}$, where m = charge, S_i is the total spin, Z = atomic number of the metal atom. Table 4.4 also shows the relevant cation–cation distances.

- It is seen that only in the case of TiO is R_E clearly less than R_C; hence it is a band system and shows the expected metallic behavior. FeO is predicted and found to be semiconducting.
- From VO onwards, the electrical conduction can occur either due to the (i) excitation of electrons from the localized t_{2g}^* or e_g^* levels to σ^* band or (ii) hopping of electrons between the metal ions. The latter is characteristic of **hopping semiconductors**.
- Doping of impurity metal ions can enhance the hopping semiconduction. A well-known example is NiO (ambient temperature $\rho = 10^{10}$–10^{12} Ω cm). The resistivity drops significantly when doped with Li^+; for example, $Li_{0.05}Ni_{0.95}O$ has $\rho = 10$–100 Ω cm. This is attributed to the presence of Ni^{3+} in the formal composition, $(Li^+)_{0.05}(Ni^{3+})_{0.05}(Ni^{2+})_{0.90}O^{2-}$, and the hopping of electrons from Ni^{2+} to Ni^{3+} (or equivalently holes from Ni^{3+} to Ni^{2+}).
- Hopping semiconductors based on ferrites show interesting conductivity trends. While Fe_3O_4 has a relatively low resistivity (ambient temperature $\rho = 0.01$–10 Ω cm), $NiFe_2O_4$ has a significantly higher resistivity ($\rho = 10^5$–10^6 Ω cm). One way to visualize the difference is in terms of the facile hopping of electrons between Fe^{2+} and Fe^{3+} sites in Fe_3O_4 $((Fe^{3+})^t \ (Fe^{2+}Fe^{3+})^o \ (O^{2-})_4$; t = tetrahedral site, o = octahedral site) and the reduced possibility of such hopping between Ni^{2+} and Fe^{3+} sites in $NiFe_2O_4$ $((Fe^{3+})^t \ (Ni^{2+}Fe^{3+})^o \ (O^{2-})_4)$, the ions at the sites being different in the latter. The low conductivity of some ferrites enables their use in transformer cores, due to the reduced eddy current losses.

- Tungsten bronzes (Section 2.4.2) also show hopping semi-conductor behavior. WO_3 is an insulator. When it is doped with A (= Li, Na, K, Rb, Cs, Mg, In, Sc, *etc.*), the resulting A_xWO_3 $(x \ll 1.0)$ contains W^{5+} in addition to W^{6+} $[(A^+)_x(W^{5+})_x(W^{6+})_{(1-x)}O_3]$. Hopping between W^{5+} and W^{6+} enhances the electrical conductivity significantly.

4.3.9 Semiconductor–Metal Transition

NiO, Fe_2O_3, *etc.*, show a semiconductor–metal transition at $\sim$6 Mbar pressure. On application of pressure, the overlap between atomic/ionic wave functions increases leading to a semiconductor–metal transition. This is known as the **Mott transition**. If the lattice constant, a (for a crystal having an odd number of electrons per primitive cell) is larger than a critical value, $a_c \sim 4.5a_0$ $(a_0 = \text{Bohr radius} = 0.53 \text{ Å})$, then the crystal is expected to be an insulator. Pressure reduces the lattice constant and transforms it into a metal.

Another interesting example of the semiconductor–metal transition is the case of V_2O_3. It shows a transition upon heating, at 150 K. At temperatures below 150 K, it is a semiconductor with $\rho = 10^2 – 10^3$ Ω cm. Above 150 K, it becomes a metal with $\rho = 10^{-3} – 10^{-5}$ Ω cm (Figure 4.24(a)). This is explained using a crystal distortion model, a band generalization of the Jahn–Teller effect. Upon cooling the crystal, the partially (1/3) filled t_{2g}^* band splits into a filled and an empty band (Figure 4.24(b)) due to lattice distortion which is stabilized at low temperature as the strain energies are small.

The Peierl's distortion leading to the metal–insulator transition in the organic charge transfer complex TTF-TCNQ at $\sim$54 K is another famous example.

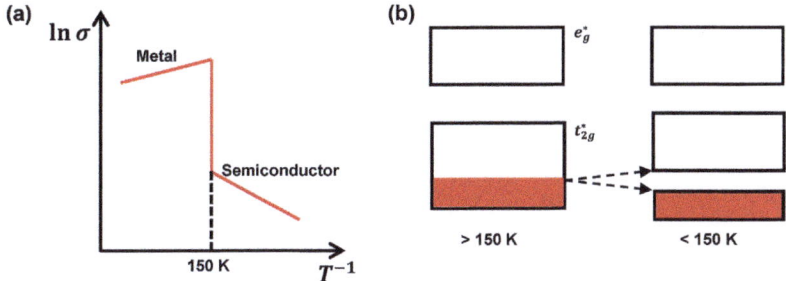

Figure 4.24 (a) Temperature (T) dependence of the electrical conductivity (σ) of V_2O_3 through the semiconductor–metal transition at 150 K. (b) Schematic diagram of the band splitting below 150 K changing the metal to a semiconductor.

4.4   Superconductivity

This phenomenon is one of the most fascinating and extraordinary properties that select materials display. Three years after he first liquefied helium, **Kamerlingh Onnes** (Leiden) reported in 1911, that the electrical resistivity of mercury dropped to a negligible value when cooled to 4.3 K and below (Figure 4.25).

- This discovery was followed by similar observations in various metals and alloys, the resistivity dropping sharply to zero, when cooled to a sufficiently low temperature, called the critical temperature, T_c.
- The material undergoes a phase transition from the **normal** to the **superconductor** state at T_c.
- In addition to several elements like Hg, Pb, Sn, In, and Tl, alloys such as Nb_3Sn and Nb_3Ge were found to become superconducting. The highest T_c among these early superconductors was 23.2 K, found in Nb_3Ge; this remained the upper limit until the discovery of the so-called 'high T_c superconductors' in the 1980s. A list of selected superconductors is given in Section 4.4.4.

4.4.1   Characteristic Features

The superconducting state formed below the **critical temperature**, T_c has several strange and unique characteristics.

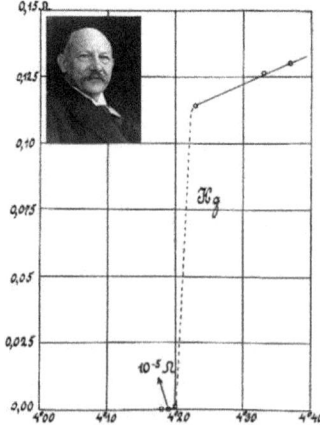

Figure 4.25   The original plot of resistance (Ω) of mercury (*y*-axis) *versus* temperature (K) (*x*-axis) by Onnes (inset: Kamerlingh Onnes), marking the discovery of superconductivity.[7,8]

- **Zero resistivity** or infinite conductivity is a defining feature of the superconducting state. Since the resistivity $\rho = \dfrac{E}{i}$, is zero for a finite current density i, the electric field, E at any point in the material is zero.
  - A consequence of this feature is that if a current is induced in a superconducting ring, the current will continue forever; this is known as the **persistent current**. Theoretical estimates suggest that the current will flow without attentuation for $10^{4.34 \times 10^7}$ s; this can be contrasted with the estimated age of the universe, 13.8 billion years $\sim 4.35 \times 10^{17}$ s.
- In 1933, Meissner and Ochsenfeld discovered that when a material enters a superconducting phase, the magnetic field is fully expelled from it; magnetic induction due to an external field or due to the superconducting current goes to zero inside the superconductor. In other words, the material exhibits **perfect diamagnetism**. This is known as the **Meissner effect**. The effect is shown schematically in Figure 4.26(a).
  - However, a sufficiently strong magnetic field can destroy superconductivity, even when the material is below its T_c. The field above which the superconducting state changes to the normal state at a specific temperature is known as the **critical field**, $H_c(T)$ at that temperature (Figure 4.26(b)).
  - The maximum current that can be passed through a superconductor at a specific temperature, T (below T_c), above which the material changes to the normal state, is called the **critical current**, $J_c(T)$ at that temperature.
  - Depending on how the magnetization, M or magnetic induction B changes with the applied magnetic field, there are two types of superconductors. **Type I superconductors** lose the

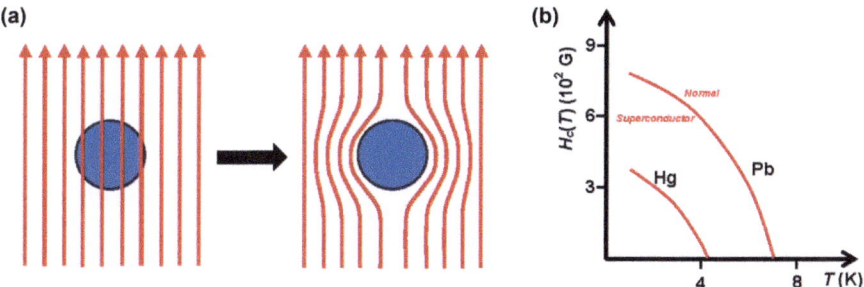

Figure 4.26 (a) Meissner effect: when the material is cooled below the T_c (indicated by the arrow), it goes into the superconducting state and the magnetic field is expelled. (b) Critical field variation with temperature for Hg and Pb; the line separates the superconducting and normal states in the H–T phase diagram.

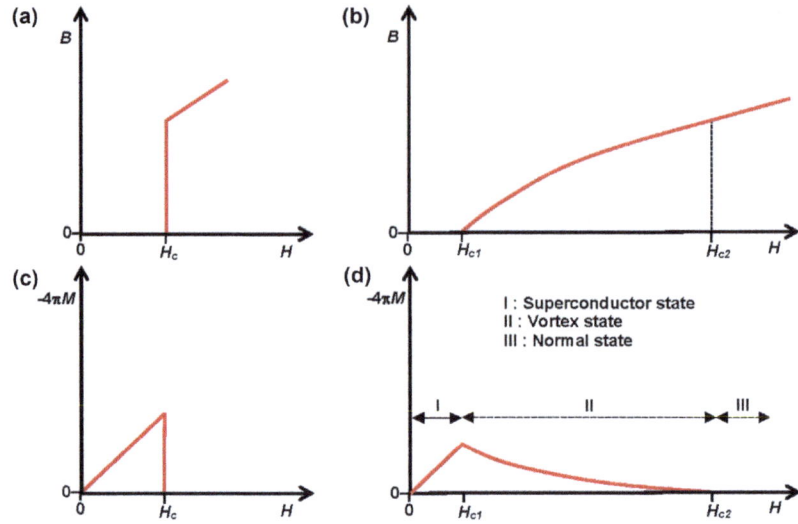

Figure 4.27 Plot of magnetic induction (*B*) *versus* magnetic field (*H*) for (a) Type I and (b) Type II superconductors below their T_c. The same information plotted as negative of the magnetization (*M*) *versus* magnetic field for (c) Type I and (d) Type II superconductors below their T_c.

superconductivity abruptly when the field exceeds H_c (Figure 4.27(a), (c)); most of the elemental superconductors such as Pb and Sn are Type I. **Type II superconductors** start losing the superconducting state gradually, when the applied magnetic field crosses the lower critical field, H_{c1}, and completely at a much higher field, the critical field, H_{c2} (Figure 4.27(b), (d)); alloys like Nb_3Sn are Type II. The region between H_{c1} and H_{c2} is known as the **vortex state**; in this state, the Meissner effect is incomplete, the magnetic field penetrates the material in vortices (magnetic flux lines). As they can sustain much higher magnetic fields, Type II superconductors are of great importance in practical applications related to high field magnets.

- In several superconductors, it is found that the T_c depends on the isotope mass, as $M^\alpha T_c = $ constant, with $0 \leq \alpha \leq 0.5$. This is known as the **isotope effect**, and indicates that superconductivity is intimately related to the lattice vibrations in these materials.
- An important characteristic of many superconductors is an **energy gap** discussed in the following section.

4.4.2 Energy Gap

Several experiments indicate the presence of an energy gap opening up at the Fermi energy, as a normal metal is transformed into a superconductor at the T_c (Figure 4.28(a)). The energy gap, $2\Delta(T)$ decreases with increasing temperature, vanishing at the T_c (Figure 4.28(b)).

- At 0 K, the gap is typically a few meV. In several superconductors, the energy gap is related to the critical temperature as, $2\Delta \sim 3.5$ $k_B T_c$.
- This energy gap is entirely different from that in an insulator/semiconductor. In an insulator/semiconductor, the gap arises due to the specific lattice and the electronic structure at the Brillouin zone edges; thus the gap is tied to the lattice. In the superconductor it is tied to the Fermi sea of electrons, and has a very different origin as discussed in Section 4.4.3. Some of the experiments which reveal the presence of the energy gap are the following.
  - Photons of energy less than $2\Delta(T)$ are not absorbed by the superconductor. But photons (typically microwave or ir) with energy higher than $2\Delta(T)$ are absorbed causing excitations across the gap.
  - AC conductivity is markedly different in the superconducting state for $\omega \leq \dfrac{\Delta}{\hbar}$.
  - Heat capacity measurements on superconductors show an exponential increase with temperature below the T_c, whereas the normal state of the same material (obtained by applying a magnetic field above H_c) shows the usual behavior expected for

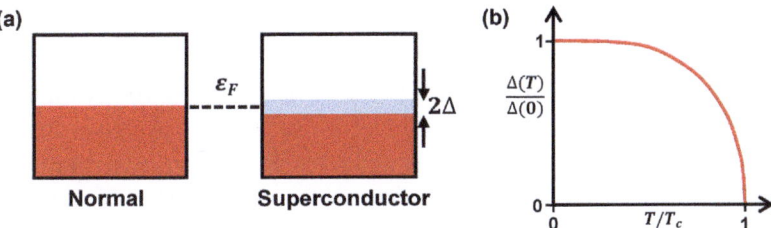

Figure 4.28 (a) Energy gap opened at the Fermi level in a superconductor. The red region is filled and the light blue region is the energy gap; sizes are not to scale, as $2\Delta \sim 10^{-4} \, \varepsilon_F$. (b) The energy gap in a superconductor decreasing with increasing temperature and vanishing at T_c.

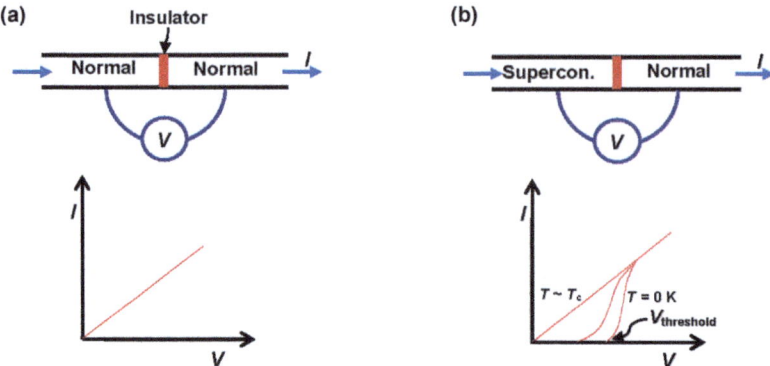

Figure 4.29 Tunnelling current across a thin insulating layer between (a) two normal metals and (b) a normal metal and a superconductor. In the latter case, the $V_{threshold}$ above which the current develops at 0 K is shown.

metals. The electronic contribution to the heat capacity shows an $e^{-1/T}$ form, characteristic of excitation across an energy gap.

○ Two metals in the normal state with an ultrathin insulating barrier in between, show simple tunnelling behavior as revealed by a linear current–voltage curve (Figure 4.29(a)). When one of the normal metals is replaced by a superconductor, the current flow arises only beyond a threshold voltage, $V_{threshold} = \dfrac{\Delta}{e}$, where e is the charge of the electron (Figure 4.29(b)).

4.4.3 Theoretical Concepts

In 1957, Bardeen, Cooper and Schrieffer (**BCS**) presented a detailed microscopic picture of the phenomenon of superconductivity. Electron pairing through the mediation of the lattice is the basic concept involved. It can be explained in a very qualitative way as follows.

- Electron–lattice–electron interaction leads to the formation of electron pairs (**Cooper pairs**) and an energy gap. An electron interacting with the positively charged lattice deforms it transiently (Figure 4.30(a)); a second electron takes advantage of this lattice deformation to lower its energy. Thus the electrons are bound into pairs through the mediation of the lattice vibrations (phonons). Electrons at the Fermi level with opposite spins, with wave vectors, $k_{F\uparrow}$, $-k_{F\downarrow}$ form the Cooper pairs (Figure 4.30(b)). The binding energy of Cooper pairs is typically of the order of meV.

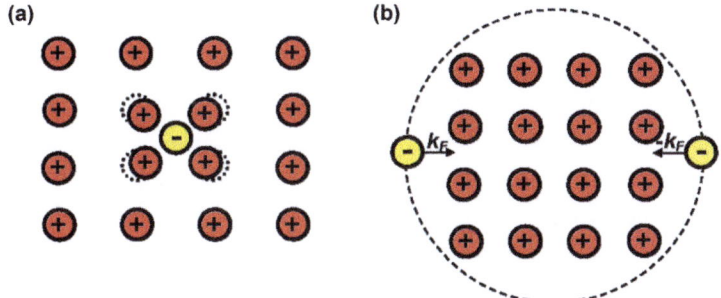

Figure 4.30 (a) Schematic diagram showing the lattice (cations) polarization due to an electron; small broken circles represent the
original positions of the ions displaced by the presence of the
electron. (b) Schematic representation of the coupling of electrons at the Fermi surface (large broken circle).

- Lower temperatures facilitate the lattice deformation; at $T > T_c$,
  the enhanced lattice vibrations lead to break-up of Cooper pairs
  and destruction of the energy gap, as well as superconductivity.
- The isotope effect arises due to the fact that a lattice of heavier
  nuclei is more difficult to deform. BCS theory provides an ex-
  pression for the critical temperature, $T_c = 1.14\theta e^{-1/U\mathcal{D}(\varepsilon_F)}$ where θ
  is the Debye temperature (takes into account the phonon
  contribution), U is the electron–lattice interaction parameter,
  and $\mathcal{D}(\varepsilon_F)$ is the density of states at the Fermi energy.
- The Cooper pairs have zero spin and are therefore bosons. Under
  an electric field, all Cooper pairs move together and are described by a single wavefunction; scattering of a single Cooper
  pair is not possible. This leads to zero resistivity.

Two characteristic lengths of superconductors are the **London
penetration depth**, λ and **coherence length**, ξ. λ measures the depth of
penetration of the magnetic field into a superconductor; this is given
by the relation, $B(x) = B_0 e^{-x/\lambda}$. ξ is a measure of the minimum spatial
extent of a transition layer between the normal and superconductor
states, the distance over which the energy gap reaches its bulk value.
Type I and II superconductors have different relative values of λ and ξ
(Figure 4.31).

High external magnetic fields increase the energy of Cooper pairs; if
it exceeds the energy gap, the pairs break up and the superconductivity is destroyed. The magnetic flux through a superconducting ring
is quantized and the effective unit of charge is $2e$, confirming the
existence of bound electron pairs.

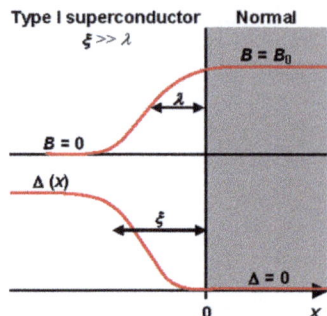 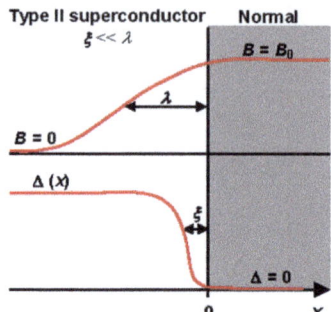

Figure 4.31 Variation of the magnetic induction (B) and energy gap parameter ($\Delta(x)$) at the interface of normal and superconductor regions (Type I and II). B_O is the magnetic induction in the normal state. Drawn schematically based on Figure 18 in Chapter 12 of ref. 2.

4.4.4 Superconducting Materials

There are several classes of materials that exhibit superconductivity at low temperatures. New families of superconductors continue to be discovered. Table 4.5 gives a list of representative superconducting materials, but this is certainly not exhaustive. The classifications are just a matter of choice.

The structure of the two organic donor molecules TMTSF and BEDT-TTF that give superconducting salts are shown in Figure 4.32. The original motivation for exploring organic superconductors came from the model proposed by Little; the idea was that a Cooper pair formed by the displacement of π-electrons of conjugated organic molecules could lead to higher T_c than those in conventional BCS superconductors involving the mediation of much heavier nuclei.

4.4.5 Superconductor Applications

Zero resistivity can obviously be exploited in power transmission without loss. However, the technical issues of maintaining temperatures below the T_c and fabricating suitable wires for transmission lines, make real-life application difficult and challenging.

Magnets based on superconducting electrical coils are used commercially in high magnetic field applications such as magnetic resonance imaging and spectrometry. In principle such magnets can be used in levitation applications for transport.

Electron pair tunnelling through insulating barriers leads to a variety of **Josephson effects** that are exploited in Josephson devices.

Table 4.5 Examples of superconducting materials.

Type	Example	T_c (K)	Remarks, if any
Element	Hg	4.2	Discovery of superconductivity
	Pb	7.2	
	Sn	3.7	
	Nb	9.5	
Alloy	Nb_3Sn	18.0	
	Nb_3Ge	23.2	Highest T_c until 1986
	V_3Ga	16.5	
	V_3Si	17.1	
Oxide	$BaPb_{0.73}Bi_{0.27}O_3$	13	
	$Ba_{1-x}K_xO_{3-y}$	29.8	
	NbO	0.9	
	$La_2CuO_4 + Ba$ (doped)	~30	First 'high T_c superconductor'
	$YBa_2Cu_3O_{7-x}$	~90	First with $T_c > $ liq. N_2 temperature
	$Tl_2Ba_2Ca_2Cu_3O_x$	125	
Sulphide	NbS_2	<6	
	TaS_2	<6	
	$A_xMo_6X_8$	<15	Chevrel phases; A = Pb, Sn, Cu, Ag, *etc.*; X = S, Se, Te
Boride	MgB_2	39	
Carbon based	KC_8, RbC_8	<1	Graphite intercalation compounds
	K_3C_{60}	~18	Fullerene based
	Cs_2RbC_{60}	33 K	Fullerene based
Pnictide	$Gd_{1-x}Th_xFeAsO$	56	
	$Ba_{1-x}K_xFe_2As_2$	38	
	$LaFeAsO_{0.9}F_{0.1}$	26	
Organic	$(TMTSF)_2PF_6$	1.4	Under 6.5 kBar
	$(BEDT\text{-}TTF)_2I_3$	8.0	α-polymorph
	$(BEDT\text{-}TTF)_2X$	12.5	$X^- = Cu[N(CN)_2Br]^-$
Polymer	$(SN)_x$	0.3	

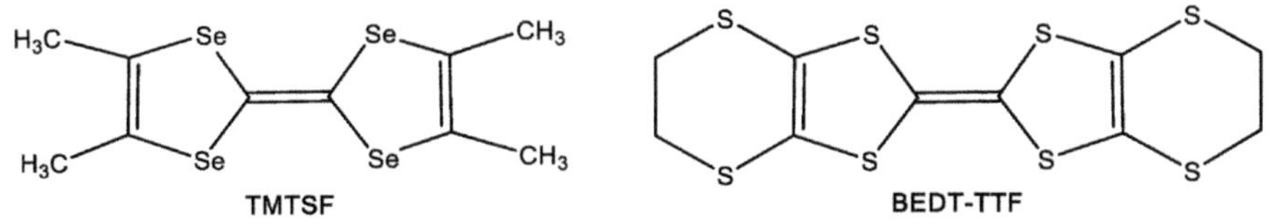

Figure 4.32 Structure of two organic π-electron donor molecules, tetramethyltetraselenafulvalene (TMTSF) and bis(ethylenedithiolato)tetrathiafulvalene (BEDT-TTF) that give superconducting salts.

A superconductor–insulator–superconductor junction enables the following effects.

- **dc Josephson effect:** A direct current can flow across the barrier, in the absence of an applied field.

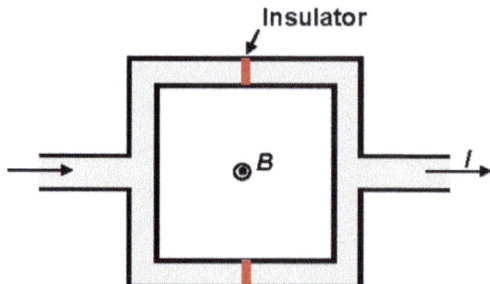

Figure 4.33   Schematic of the arrangement for the macroscopic quantum interference; the grey region is the superconductor. *B* represents the magnetic field which gives rise to the flux, φ through the interior of the loop, and *I* is the current.

- **ac Josephson effect**: In the presence of a dc voltage, V applied across the junction, an rf current oscillation occurs with frequency, $\omega = \dfrac{2eV}{\hbar}$. This has been used to make precise determination of the universal constant, $e/\hbar$.

- **Macroscopic quantum interference**: A dc magnetic field applied through a circuit containing two junctions in parallel (Figure 4.33), causes the maximum supercurrent to show interference effects dependent on the magnetic field intensity. The current is given by:

$$I = I_0 \left| \frac{\sin\left(\dfrac{\pi\varphi}{\varphi_0}\right)}{\left(\dfrac{\pi\varphi}{\varphi_0}\right)} \right|,$$ where φ is the total magnetic flux at the junction

and $\varphi_0 = \dfrac{hc}{2e}$. The current oscillation allows a sensitive determination of the external magnetic field. This principle is used in the design of superconducting quantum interference device (SQUID) magnetometers.

4.5   Dielectrics: Piezo, Pyro and Ferroelectrics

Dielectrics are essentially electrically insulating materials that show a polarization (also called dielectric polarization, the net charge displacement or induced dipole moment per unit volume) when subjected to an electric field. A detailed discussion of dielectric polarization resulting from the interaction of materials with the oscillating electric field of an electromagnetic radiation is presented in Section 6.1.2. The important phenomena associated with the effect of an external electric field, as well as the impact of mechanical stress and temperature on the polarization of dielectrics, are highlighted here. Crystallographic symmetry of the materials is an important aspect in this context.

- Polarization (*P*) of a dielectric is proportional to the applied electric field (*E*), at relatively low fields; nonlinear dependence at higher fields is discussed in Section 6.6.
- The linear behavior can be represented by the equation, $P = \chi E$.
- χ is the dielectric susceptibility (also called electric susceptibility); since *P* and *E* are vectors, χ is a second-rank tensor, and the equation is more appropriately written as $P_i = \Sigma_j \chi_{ij} E_j$ considering the components of each quantity.
- Figure 4.34 shows the linear behavior of the polarization, and a schematic representation of the charge displacements that lead to the polarization.

4.5.1 Piezoelectric Materials

Dielectric polarization occurs in any material irrespective of the symmetry of the crystal lattice, the extent of polarization in a given electric field being determined by the dielectric susceptibility. Several materials with a non-centrosymmetric lattice, exhibit a polarization under mechanical stress; these are called piezoelectric materials.

- The piezoelectric coefficient, *d* relates the polarization with the stress, σ (the pressure experienced per unit area of the material). The relevant equation is: $P = d\sigma$. As *P* is a vector and σ a second-rank tensor, *d* is in general a third-rank tensor.
- One can also define a converse piezoelectricity, as the development of a strain (ε) in a material in the presence of an electric field, $\varepsilon = dE$. ε is the deformation of the material, usually measured in terms of the relative change in length, volume, *etc.*,

As noted earlier, symmetry is an important criterion which determines whether piezoelectric and related effects can potentially be observed in a crystal. Therefore, a brief look at the aspects of crystal symmetry is useful.

- The geometrical representation of any physical property of a crystal follows the symmetry of the point group of the crystal; this is known as **Neumann's principle**. In other words, if the crystal

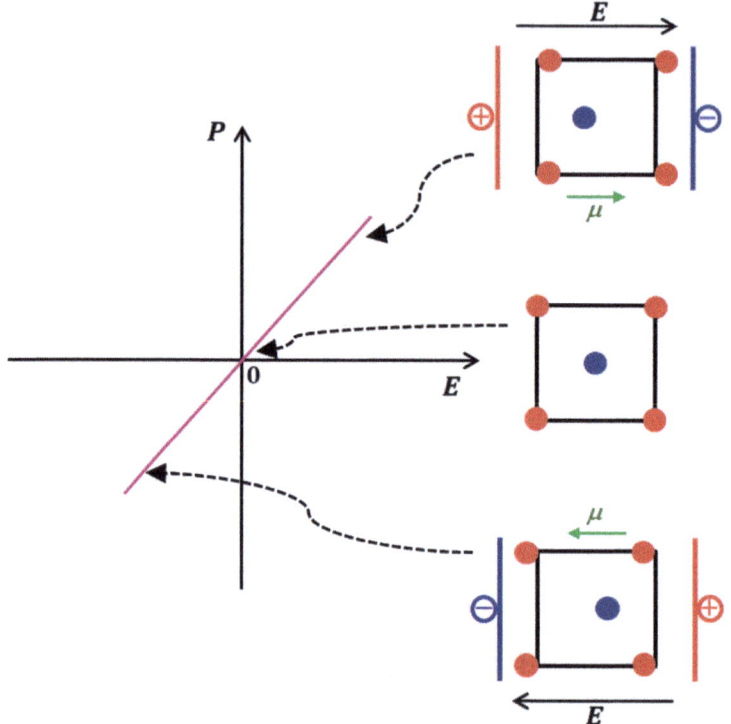

Figure 4.34 Plot of the linear variation (purple line) of the polarization (P) of a dielectric with the applied electric field (E). Schematic representation of the positive (red) and negative (blue) ions in the unit cell of a crystal with net zero dipole moment (when $E = 0$), and their relative displacements leading to dipole moment (μ) generation in opposite directions in response to the change in direction of E.

possesses certain symmetry elements, its physical properties must also be invariant under these symmetry operations.

- Among the 32 crystallographic point groups (Section 1.3.6), 21 are non-centrosymmetric (lacking a center of inversion symmetry element). Among the latter, 10 are polar groups (having only a single rotation axis).
- A summary of the non-centrosymmetric point groups is provided in Table 4.6.

Piezoelectricity is possible only in crystals that belong to a non-centrosymmetric point group.

- This follows from the fact that all elements of odd-rank tensors vanish in an inversion-symmetric system.

Table 4.6 Non-centrosymmetric crystal point groups under each of the seven crystal systems; those which are polar (pyroelectric materials belong to these groups) are indicated in bold font, and the only non-centrosymmetric group which does not support piezoelectricity is O (432), underlined. In all cases, Schoenflies symbols are shown with the corresponding Hermann–Mauguin notation in parenthesis.

Triclinic	Monoclinic	Orthorhombic	Tetragonal	Trigonal	Hexagonal	Cubic
C_1 **(1)**	C_2 **(2)**	C_{2v} **(mm2)**	C_4 **(4)**	C_3 **(3)**	C_6 **(6)**	T (23)
	C_s **(m)**	D_2 (222)	C_{4v} **(4 mm)**	C_{3v} **(3m)**	C_{6v} **(6 mm)**	T_d ($\bar{4}$3m)
			S_4 ($\bar{4}$)	D_3 (32)	C_{3h} ($\bar{6}$)	$\underline{O}$ (432)
			D_4 (422)		D_6 (622)	
			D_{2d} ($\bar{4}$2m)		D_{3h} ($\bar{6}$2m)	

- Figure 4.35 presents schematic examples to visualize in a simple way how mechanical stress produces a polarization when a center of inversion is absent (T_d point group), but does not when a center of inversion is present (O_h point group); note that T_d appears in Table 4.6, but not O_h.
- The O group (432) is the lone non-centrosymmetric one that does not support piezoelectricity, due to the specific combination of symmetry elements present in it.

Quartz is a classic example of a piezoelectric material; an important use is that of a quartz crystal oscillator in clocks. Application of an alternating electric field causes structural oscillations of the crystal; the precise frequency of the oscillation (typically 2^{15} Hz) can be used to monitor time. Other well-known piezoelectric materials are $BaTiO_3$, $LiNbO_3$, and ZnO. Piezoelectric materials are used in various applications including sensors and actuators, where the interplay of electrical and mechanical effects is exploited.

Electrostriction is also a phenomenon in which a structural strain is induced in a material upon application of an electric field. However, this can occur in any dielectric material (irrespective of its symmetry) and is due to the displacement of ions. The strain in this case is proportional to the square of the electric field.

4.5.2 Pyroelectric and Ferroelectric Materials

As noted in Table 4.6, 10 of the crystallographic point groups are non-centrosymmetric and also polar. They possess a dipole moment even in the absence of an external electric field or mechanical stress; this is known as **spontaneous polarization**.

Pyroelectric materials belong to these polar crystal point groups; they are characterized by a polarization change with temperature.

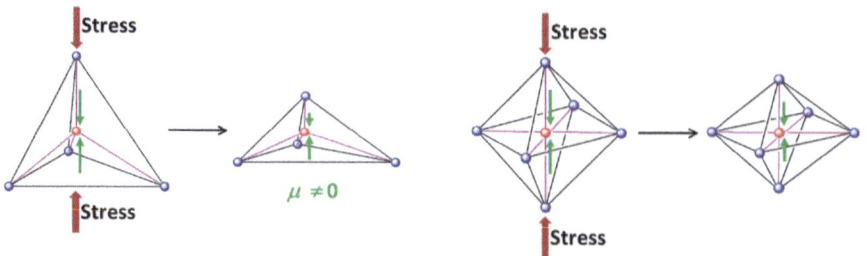

Figure 4.35   Tetrahedral and octahedral arrangement of anions (blue) around a cation (red), and illustration of the generation of a non-zero dipole moment (green arrow) upon application of a stress in the former, but not in the latter.

- The spontaneous polarization changes due to minute changes in the atomic/ionic positions when the temperature is varied.
- Change in the polarization on the sample surface can be measured as an induced electric current.
- The pyroelectric coefficient, p relates the change in polarization with the change in temperature, $\Delta P = p\Delta T$. Since polarization (and its change) is a vector and temperature (and its change) a scalar, the pyroelectric coefficient is a vector.

In select pyroelectric materials, application of an electric field and its reversal, leads to switching of the direction of polarization; such pyroelectrics are called **ferroelectric materials**.

- The spontaneous polarization and its reversal can be observed only below a phase transition (Curie) temperature, T_C.
- The hysteresis behavior in the polarization *versus* temperature, observed at $T < T_C$ is shown in Figure 4.36(a); the analogous plot of magnetization *versus* temperature is explained in detail in Section 5.4.
- The difference between ferroelectric and other pyroelectric materials can be seen in the schematic diagram of the ionic displacements in the two cases at temperatures below and above T_C, in response to an alternating electric field (Figure 4.36(b)).

Ferroelectric materials have domain structures with their spontaneous polarization vectors oriented randomly. Electric field poling can be used to align them. Phenomena such as antiferroelectricity are also known. Conceptually, these are very similar to the various aspects

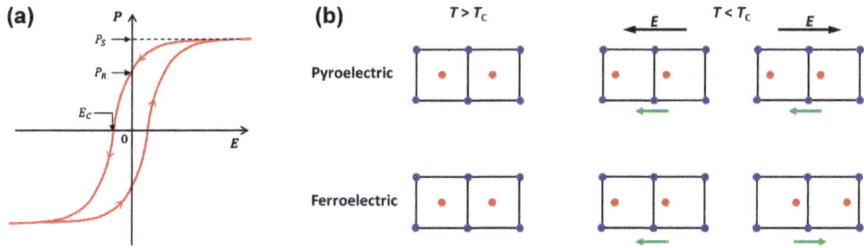

Figure 4.36   (a) Plot of the polarization (P) *versus* electric field (E) for a ferroelectric at $T < T_C$; saturation polarization (P_S), remanent polarization (P_R) and coercive field (E_C) are indicated. (b) Schematic figure of a pair of unit cells in pyroelectric and ferroelectric crystals, indicating the placement of anions (blue) and cations (red) and their relative displacements above and below T_C, under alternate directions of electric field (E) in the latter cases; dipole moments are indicated by green arrows.

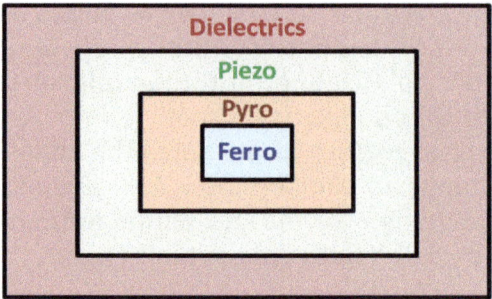

Figure 4.37   Hierarchy of dielectric materials, piezo, pyro and ferroelectrics.

of ferromagnetism and related magnetic phenomena discussed in detail in Section 5.4.

Examples of pyroelectrics and ferroelectrics include perovskites like $LiTaO_3$ and triglycine sulphate. An interesting ferroelectric polymer is poly(vinylidine fluoride); the combination of ferroelectricity and polymer flexibility is unique. An important application of pyroelectric materials is in thermal sensing. Ferroelectric thin films are used in non-volatile memory applications and micro-electromechanical systems.

An overview of the hierarchy of dielectrics together with the piezo, pyro and ferroelectrics is indicated in Figure 4.37. All ferroelectrics can exhibit piezo and pyroelectric effects; similarly all pyroelectrics are also piezoelectric. The converse is not true.

References

1. https://www.doitpoms.ac.uk/tlplib/semiconductors/fermi.php
2. C. Kittel, *Introduction to Solid State Physics*, Wiley, New York, 5th edn, 1976.
3. M. Whangbo, in *Extended Linear Chain Compounds*, ed. J. S. Miller, Plenum Press, New York, 1982, p. 127.
4. G. I. Epifanov, *Solid State Physics*, Mir Publishers, Moscow, 1979.
5. http://hypertextbook.com/facts/2004/JennelleBaptiste.shtml
6. J. B. Goodenough, *Annu. Rev. Mater. Sci.*, 1971, **1**, 101.
7. https://commons.wikimedia.org/wiki/File:Superconductivity_1911.png#filelinks
8. https://commons.wikimedia.org/wiki/Heike_Kamerlingh_Onnes#/media/File:Kamerlingh_portret.jpg

5 Magnetic Properties

5.1 Classification of Magnetic Materials

There is great variety in magnetic phenomena and magnetic materials. Before looking more deeply into some of the most common ones, an overview of the classification of magnetic materials is useful. Magnetic moment of an atom arises due to the following:

- spin of the electrons
- orbital angular momentum of the electrons
- change in orbital moment of electrons induced by an applied magnetic field
- spin of the nucleus

The first three, due to the electrons, are the dominant effects; the first two provide a paramagnetic contribution (positive susceptibility), while the third gives a diamagnetic contribution (negative susceptibility) to the magnetic moment (these are discussed in the following sections, and for definitions of the basic terms, see Section 5.6). Magnetic moment due to the nuclear spin is very small compared to that due to the electrons, as nuclei have much smaller gyromagnetic ratio (due to the significantly higher mass).

5.1.1 Paramagnetism and Diamagnetism

The most basic characterization of the magnetic behavior of materials is in terms of para- and diamagnetism. If an atom, molecule, or material is attracted into a magnetic field, it is said to be **paramagnetic**,

Core Concepts for a Course on Materials Chemistry
By T. P. Radhakrishnan
© T. P. Radhakrishnan 2023
Published by the Royal Society of Chemistry, www.rsc.org

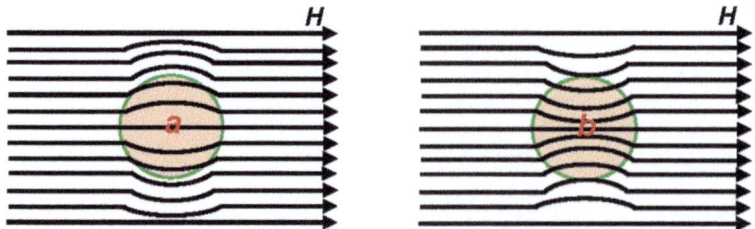

Figure 5.1 Magnetic lines of force in the presence of a (a) diamagnetic and (b) paramagnetic body.

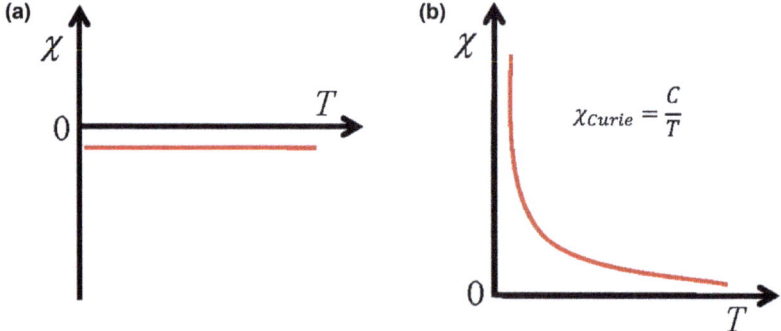

Figure 5.2 Schematic plot of magnetic susceptibility (χ) *versus* temperature (T) for (a) diamagnetic and (b) paramagnetic samples; the scales for the χ axis are very different in the two cases. The plot in (b) follows the Curie law shown on the graph.

and if repelled by the magnetic field, it is designated as **diamagnetic**. The magnetic lines of force are pulled into the paramagnetic substance, but pushed out of a diamagnetic substance (Figure 5.1).

In the presence of a magnetic field, **H**, paramagnetic materials show a magnetization, **M** parallel to **H**, whereas diamagnetic materials show a magnetization, **M** antiparallel to **H**. Consequently, paramagnetism is associated with a positive susceptibility, and diamagnetism with a negative susceptibility. Examples are:

- diamagnetic materials: $H_2(g)$, H_2O, $NaCl$, C_6H_6
- paramagnetic materials: $O_2(g)$, $CuSO_4 \cdot 5H_2O$, 2,2-diphenyl-1-picrylhydrazyl (DPPH)

Temperature dependence of diamagnetism and paramagnetism are very different. Diamagnetic susceptibility is generally very small and temperature independent, whereas paramagnetic susceptibility is significantly larger and in most systems, strongly temperature dependent (Figure 5.2).

5.1.2   Ferro-, Antiferro-, and Ferrimagnetism

Paramagnetism is characterized by the presence of non-interacting magnetic moments of individual species in the material. Cooperative interactions between magnetic moments of atoms, ions, or molecules in a material give rise to a variety of magnetic phenomena. The most common ones are ferro-, antiferro- and ferrimagnetism.

- Materials in which the neighboring magnetic moments tend to align parallel are ferromagnetic.
- Materials in which the neighboring magnetic moments tend to align antiparallel are antiferromagnetic.
- Materials in which the neighboring magnetic moments of different magnitude tend to align antiparallel, resulting in a net moment, are ferrimagnetic. Ferrimagnetism is known also in materials with an odd number of moments in the unit cell.

After discussing the origin of diamagnetism (Section 5.2) and para-magnetism (Section 5.3), the cooperative effects will be discussed in Section 5.4. It must be noted that there are a number of other cooperative magnetic phenomena such as metamagnetism, heli-magnetism, canted antiferromagnetism, *etc.*, which are rather special-ized cases.

5.2   Diamagnetism

Lenz's law in electromagnetism shows that when the magnetic flux through an electric circuit is changed, it induces a current in such a way as to oppose the flux change. In other words, the magnetic field due to the induced current opposes the applied field; the magnetic moment associated with the induced current is a **diamagnetic moment**.

Diamagnetism of atoms, molecules, and materials arises due to the change in orbital motion of electrons due to the applied magnetic field. This effect is prominent in materials with a zero atomic mag-netic moment; it is overwhelmed by the larger paramagnetic moment, if present.

5.2.1   Larmor Frequency

The motion of electrons around a central nucleus in the presence of a magnetic field is the same as a possible motion in the absence of the field, except for the superposition of a precession frequency; this is the Larmor theorem, and the precession frequency is called the **Larmor frequency**. It can be visualized as follows.

- In the absence of any magnetic field, the centripetal force on a moving electron due to the nuclear charge is $F = \dfrac{mv_0^2}{r} = m\omega_0^2 r$ (where, $m =$ mass of the electron, $r =$ radius of the orbit, v_0, $\omega_0 =$ linear, angular velocity respectively).
- Electrons with opposite angular velocity and hence opposite angular momentum, will produce magnetic moment in opposite directions. (Note: magnetic moment, μ is related to the angular momentum, p by the gyromagnetic ratio, γ: $\vec{\mu} = \gamma \cdot \vec{p}$). The direction of the magnetic moment is given by the right hand thumb rule (note that the direction of current is opposite to the direction of motion of electrons).
- Figure 5.3(a) and (b) show the situation in the absence of the external magnetic field, with the net magnetic moment zero.
- In the presence of an external magnetic field, B, a Lorentz force $(F_B = -ev \times B$, where e is the charge of the electron) appears. In Figure 5.3(c), $v \times B$ points outwards and hence the direction of F_B is inward, adding to the centripetal force, F. As a result of the increased centripetal force, $(F + F_B)$, the electron is accelerated resulting in an increased magnetic moment, shown as $(\mu + \Delta\mu)$. Similarly, in Figure 5.3(d), it is seen that the electron with the opposite orbital motion is decelerated and gives rise to a decreased magnetic moment, $(\mu - \Delta\mu)$ in the opposite direction. Net increase in the magnetic moment is $2\Delta\mu$ in a direction opposite to that of B. This is the origin of diamagnetism.

The increased/decreased centripetal forces shown in Figure 5.3(c) and (d) can be written as follows, and related to the increased/decreased angular velocities as:

$$F_1 = m\omega_0^2 r + ev_0 B = m\omega_0^2 r + e\omega_0 rB = m\omega_1^2 r$$

$$F_2 = m\omega_0^2 r - ev_0 B = m\omega_0^2 r - e\omega_0 rB = m\omega_2^2 r$$

- Subtracting the second equation from the first,

$$m(\omega_1^2 - \omega_2^2)r = mr\,(\omega_1 + \omega_2)(\omega_1 - \omega_2) = 2e\omega_0 rB$$

- **Larmor frequency**, ω_L is the precession frequency superimposed on the electron in presence of the magnetic field. $\omega_1 = \omega_0 + \omega_L$ and $\omega_2 = \omega_0 - \omega_L$

$$\therefore (\omega_1 + \omega_2) = 2\omega_0 \text{ and } (\omega_1 - \omega_2) = 2\omega_L$$

- Finally, $mr(\omega_1 + \omega_2)(\omega_1 - \omega_2) = 4mr\omega_0\omega_L = 2e\omega_0 rB \Rightarrow \omega_L = \dfrac{eB}{2m}$

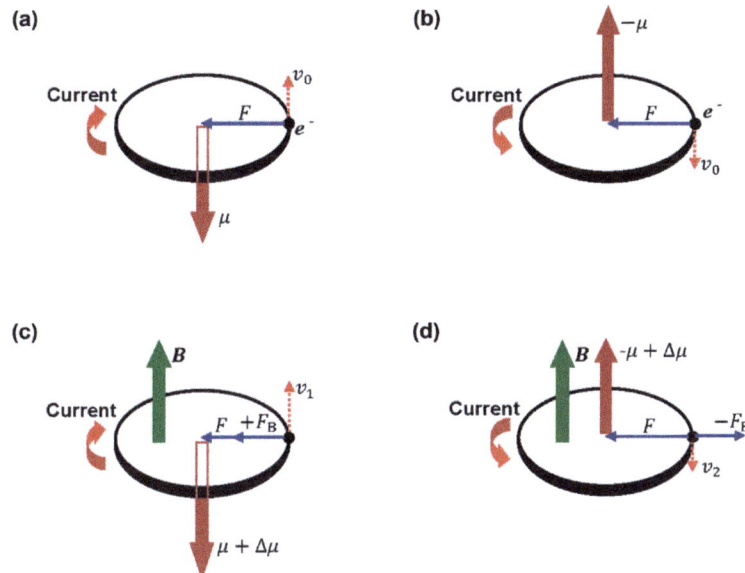

Figure 5.3 Schematic diagram showing the opposite orbital motion of a pair of electrons (v is the velocity) and the corresponding directions of the current, the centripetal force (F), and the resulting magnetic moment, μ in each case; v and F are in the plane of the orbit, whereas μ is perpendicular to it: (a, b) in the absence of an external magnetic field, and (c, d) in presence of the external field, B. In the latter, increase and decrease of the velocities (from v_0 to v_1 and v_2) and the centripetal force ($F+F_B$ and $F-F_B$) and corresponding increase and decrease of the magnetic moments antiparallel and parallel to B respectively are indicated.

5.2.2 Langevin Diamagnetic Susceptibility

The Larmor precession frequency introduced due to the presence of the external field, B, leads to a current around the nucleus. The induced magnetic moment due to this current opposes the applied field as shown above.

- The induced current, $I = q \times \nu$, where q is the total charge of particles and ν is the number of revolutions per unit time.
  - For an atom with atomic number Z, $q = -Ze$; $\nu = \dfrac{\omega_L}{2\pi}$
  - $I = -Ze \times \dfrac{1}{2\pi}\left(\dfrac{eB}{2m}\right) = \dfrac{-Ze^2B}{4\pi m}$
- Magnetic moment induced, $\mu = I \times A$, where A is the area of the loop (orbit).
- $A = \pi \langle \rho^2 \rangle$, where $\langle \rho^2 \rangle = (\langle x^2 \rangle + \langle y^2 \rangle)$ is the mean square of the perpendicular distance of the electron from the field axis through the nucleus.
  - The mean square distance of the electron from the nucleus in 3-D is given by, $\langle r^2 \rangle = (\langle x^2 \rangle + \langle y^2 \rangle + \langle z^2 \rangle)$

Table 5.1   Diamagnetic susceptibility (in CGS units) of some elements.

Atom	Bi	Ge	Si	He	Ne	Ar	Kr
χ_{dia} $(10^{-6}\ \text{cm}^3\,\text{mol}^{-1})$	−18.0	−0.8	−0.3	−1.9	−7.2	−19.4	−28.0

- For a spherically symmetric distribution of charge, $\langle x^2 \rangle = \langle y^2 \rangle = \langle z^2 \rangle$; then, $\langle \rho^2 \rangle = \dfrac{2}{3}\langle r^2 \rangle$

- From the above considerations, $\mu = I \times A = \dfrac{-Ze^2 B}{4\pi m} \times \dfrac{2\pi}{3}\langle r^2 \rangle$
$$= \frac{-Ze^2 B \langle r^2 \rangle}{6m}$$

- For a material with N atoms per unit volume, the magnetization, the magnetic moment per unit volume,
$$M = \frac{-NZe^2 B \langle r^2 \rangle}{6m}$$

- Since susceptibility is given by, $\chi = \dfrac{M}{H} = \dfrac{\mu_0 M}{B}$, the diamagnetic susceptibility is found to be, $\chi_{\text{dia}} = -\dfrac{\mu_0 N Z e^2 \langle r^2 \rangle}{6m}$ (SI units)

- The above expression clearly shows the main characteristics of diamagnetic susceptibility, the negative sign (consistent with Figure 5.3) and temperature independence (see also Figure 5.2(a)).

- The magnitude of the diamagnetic susceptibility is usually rather small; typical values of some elements are shown in Table 5.1.

5.3   Paramagnetism

Electronic paramagnetism leads to a positive susceptibility and occurs commonly in:

- Atoms, molecules, solids with lattice defects, *etc.*, with an odd number of electrons, for example: Na atom, NO(g), organic free radicals like DPPH, F-center in alkali halides.
- Free metal ions with partially filled inner shells, for example: Mn^{2+}, Gd^{3+}, U^{4+}, and also in many of their complexes.
- Some molecules with an even number of electrons, for example, O_2(g), and organic diradicals.
- Metals with free electrons.

Paramagnetic susceptibility in all systems except the last one is temperature dependent. The dependence is described by Curie law.

5.3.1 Quantum Theory of Paramagnetism

The magnetic moment of a free atom/ion, $\mu = \gamma \hbar J = -g\beta J$, where $\hbar J$ is the total angular momentum [orbital $(\hbar L)$ + spin $(\hbar S)$], γ is the gyromagnetic ratio, $\beta = \dfrac{e\hbar}{2mc}$ is the Bohr magneton (in CGS units), and $g = 1 + \dfrac{J(J+1) + S(S+1) - L(L+1)}{2J(J+1)}$ is the Landé factor with a value of 2.0023 (often approximated as 2.0) for an electron with spin and orbital angular momenta.

- Energy levels of the system in a magnetic field, H are given by:
  - $U = \mu \cdot H = m_J g \beta H$, where m_J quantum number has the values, J, $J-1, \ldots 0, \ldots -J+1, -J$.

- Consider the simplest case of a spin 1/2 system with no orbital moment:
  - $m_J = m_S = \pm\frac{1}{2}$
  - Since $g \cong 2.0$, $U = \pm\beta H$. If the magnetic field is along the positive z-axis, the corresponding magnetic moments, $\mu_z = \mp\mu$, where $\mu = g\beta|m_S| \cong \beta$.
  - The energy level splitting and the corresponding magnetic moments are shown in Figure 5.4(a). Since $\mu = -g\beta S$, the magnetic moment of an electron is opposite in sign to the spin, S. The lower energy state corresponds to the magnetic moment being parallel to the magnetic field.
  - The populations of the lower (N_1) and higher (N_2) energy levels are given by the Boltzmann distribution:

$$\frac{N_1}{N} = \frac{\exp\left(\dfrac{\mu H}{k_B T}\right)}{\exp\left(\dfrac{\mu H}{k_B T}\right) + \exp\left(\dfrac{-\mu H}{k_B T}\right)} \quad \text{and} \quad \frac{N_2}{N} = \frac{\exp\left(\dfrac{-\mu H}{k_B T}\right)}{\exp\left(\dfrac{\mu H}{k_B T}\right) + \exp\left(\dfrac{-\mu H}{k_B T}\right)}$$

  - The fractional population changes with temperature as shown in Figure 5.4(b).
  - As the lower level with the higher population has a magnetic moment parallel to the external field, the net magnetization is aligned along the external field. Figure 5.4(a) shows that the resultant magnetization, M for N atoms per unit volume is:

$$M = N_1\mu + N_2(-\mu) = (N_1 - N_2)\mu = N\mu \cdot \frac{e^x - e^{-x}}{e^x + e^{-x}} = N\mu \cdot \tanh(x),$$

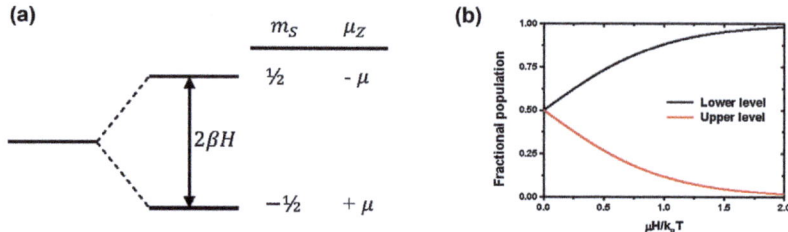

Figure 5.4 (a) Energy level splitting for one electron in a magnetic field H; the quantum number (m_s) and magnetic moment (μ_z) of each level, and the energy gap ($2\beta H$) are indicated. (b) The corresponding fractional population in thermal equilibrium at temperature T. Drawn schematically based on Figures 2 and 3 in Chapter 14 of ref. 1.

where $x = \dfrac{\mu H}{k_B T}$

○ Under the conditions, $\mu H \ll k_B T$, or $x \ll 1$, $\tanh(x) \cong x$, and

$$M = N\mu \left(\frac{\mu H}{k_B T} \right).$$

- Now consider an atom with angular momentum quantum number J. In the presence of a magnetic field, it will split into $2J + 1$ equally spaced levels. The magnetization is then given by:

○ $M = NgJ\beta B_J(x)$, where $x = \dfrac{gJ\beta H}{k_B T}$, and $B_J(x)$ is called the Brillouin function, defined as:

$$B_J(x) = \frac{2J+1}{2J} \mathrm{ctnh}\left(\frac{(2J+1)x}{2J} \right) - \frac{1}{2J} \mathrm{ctnh}\left(\frac{x}{2J} \right)$$

○ Since $\mathrm{ctnh}(x) = \dfrac{1}{x} + \dfrac{x}{3} - \dfrac{x^3}{45} + \cdots$, for $x \ll 1$, $\mathrm{ctnh}(x) \cong \dfrac{1}{x} + \dfrac{x}{3}$, the Brillouin function can be simplified as:

$$B_J(x) = \frac{2J+1}{2J} \left\{ \left(\frac{2J}{(2J+1)x} \right) + \left(\frac{(2J+1)x}{6J} \right) \right\}$$

$$-\frac{1}{2J} \left\{ \frac{2J}{x} + \frac{x}{6J} \right\} = \{(2J+1)^2 - 1\} \frac{x}{12J^2} = \frac{J(J+1)x}{3J^2}$$

○ This leads to: $M = NgJ\beta \left(\dfrac{J(J+1)x}{3J^2} \right) = NgJ\beta \left(\dfrac{J(J+1)}{3J^2} \right) \left(\dfrac{gJ\beta H}{k_B T} \right)$

$$\therefore M = \frac{NJ(J+1)g^2\beta^2 H}{3k_B T}$$

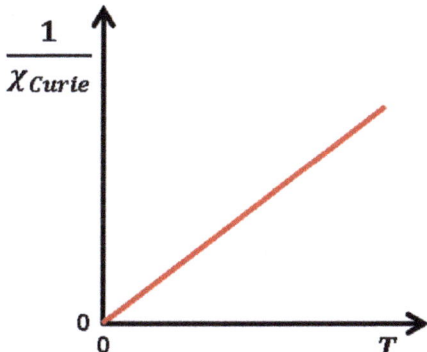

Figure 5.5 Plot of inverse magnetic susceptibility *versus* temperature for a Curie paramagnet.

- Brillouin function plot (*M versus H/T*) is often employed to obtain direct estimation of the angular momentum quantum number *J* of the atom/ion.
  - The paramagnetic susceptibility, $\chi = \dfrac{M}{H} = \dfrac{Np^2\beta^2}{3k_\mathrm{B}T}$, where $p = g\sqrt{J(J+1)}$ is known as the effective Bohr magneton.
  - The above equation gives the **Curie law**, written as $\chi_\mathrm{Curie} = \dfrac{C}{T}$, where *C* is the Curie constant (Figure 5.2(b)).
  - The characteristic plot of inverse susceptibility *versus* temperature for Curie paramagnets is shown in Figure 5.5. Note that it goes through the origin, showing that the susceptibility diverges when the temperature goes to zero.

5.3.2 Pauli Paramagnetism

Most metals show a weak and temperature-independent paramagnetism. The paramagnetic susceptibility of the conduction electrons is known as **Pauli susceptibility**. *N* electrons with a magnetic moment of one Bohr magneton are expected to contribute to a Curie type paramagnetism with the magnetization given by,

$$M = \frac{N\beta^2 H}{k_\mathrm{B}T}.$$

However the magnetization of most non-ferromagnetic metals is found to be independent of temperature.

- The result shown above for free electrons assumes (as shown in Section 5.3.1) that, the probability of an electron moment being aligned parallel to an external magnetic field exceeds the probability of the moment being aligned antiparallel, by $\sim \dfrac{\beta H}{k_\mathrm{B}T}$.

- However, as shown in Section 4.2, the electrons in metals are not free particles; they occupy energy bands.
- In the absence of an external magnetic field, the conduction band can be represented in the form of two half-bands with equal numbers of electrons with opposite spin moments (Figure 5.6(a)). The total magnetic moment is zero.
- When an external field is applied, the half-band with spin moments parallel to the field is stabilized by βH and the other destabilized by an equal energy (Figure 5.6(b)).
- The shift in the energies is small, as the magnetic field is a very weak perturbation. In order to ensure a common Fermi energy at equilibrium, electrons from the higher band flip over to the lower band. The fraction of electrons that can do this at temperature, T is $\sim \dfrac{T}{T_F}$ where T_F is the Fermi temperature. This leads to equalization of Fermi energy in the two half-bands (Figure 5.6(c)).
- The resulting net magnetization is now determined by $N' \sim N\left(\dfrac{T}{T_F}\right)$ and not N electrons.
- Therefore, $M \sim N\left(\dfrac{T}{T_F}\right) \cdot \dfrac{\beta^2 H}{k_B T} = \dfrac{N\beta^2 H}{k_B T_F}$. A detailed calculation taking into account, the Fermi–Dirac distribution gives,

$$M = \frac{3N\beta^2 H}{2k_B T_F}.$$

- Thus, the Pauli paramagnetic susceptibility is given by:

$$\chi_{Pauli} = \frac{3N\beta^2}{2k_B T_F} = \frac{3N\beta^2}{2\varepsilon_F}, \text{where, } \varepsilon_F \text{ is the Fermi energy of the metal.}$$

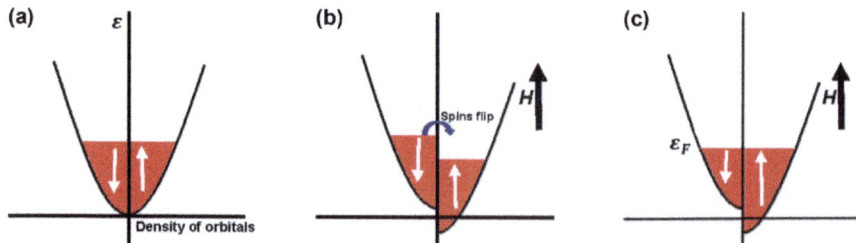

Figure 5.6 Conduction band of a metal with spin-up and spin-down half bands (a) in the absence of an external magnetic field, and (b, c) in the presence of an external magnetic field, H showing the spin flips that ensure the same chemical potential (Fermi level, ε_F) for the two half-bands, resulting in a net magnetization parallel to the external field (paramagnetism); note that only a small fraction of the total electrons, near the ε_F are involved in the flip.

Landau diamagnetism arises from the quantization of the energy of conduction electrons and the change in the energy in the presence of a magnetic field; trajectories of the electrons bend in the presence of the field $\chi_{\text{Landau}} \sim -\frac{1}{3}\chi_{\text{Pauli}}$.

5.3.3 Adiabatic Demagnetization

Temperatures well below 1 K (typically 10^{-3} K) can be reached using the method of adiabatic demagnetization. The principle involved is the following: an adiabatically sealed lattice cools when the entropy associated with the distribution of spin moments of a paramagnetic system, increases due to the removal of the ordering influence of a magnetic field.

The process can be visualized as follows (Figure 5.7).

- A paramagnetic system at finite temperature, in the absence of a magnetic field, possesses spin entropy resulting from the random distribution of the spin moments.
- When a magnetic field is imposed on the sample under isothermal condition (in good thermal contact with the surrounding, provided by helium gas), the spin level degeneracies split, the spins are partly ordered, and the entropy reduced

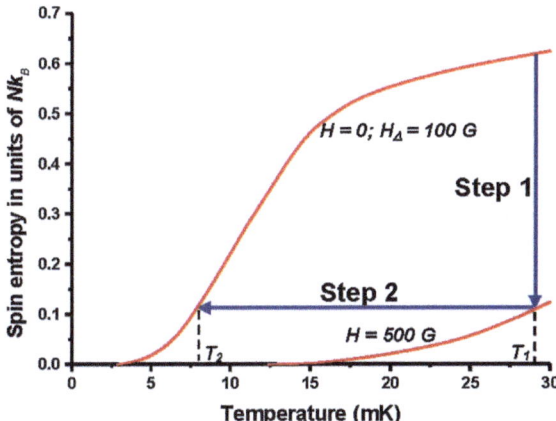

Figure 5.7 Schematic plot showing the entropy of an $S = 1/2$ system as a function of temperature. Top red curve: with no external magnetic field, but only an internal magnetic field H_Δ; bottom red curve: in the presence of a large external magnetic field. Step 1: magnetization under isothermal condition; Step 2: demagnetization under adiabatic condition. Initial and final temperatures, T_1 and T_2 are indicated. Drawn schematically based on Figure 8 in Chapter 14 of ref. 1.

(Step 1 in Figure 5.7). The entropy is also lowered if the temperature is decreased as more of the moments line up.

- If the magnetic field is now removed keeping the sample in an adiabatic condition (removing all thermal contact with the surrounding by pumping out the helium gas), the total entropy remains constant, but the spin system looks like one at a lower temperature (Step 2 in Figure 5.7); equivalently, entropy flows into the spin system from the lattice. The process can be visualized as the lattice entropy decreasing and the spin entropy increasing (Figure 5.8); note that the lattice entropy should be very low (*i.e.*, low starting temperature) compared to the spin entropy. Another way to visualize the effect is, the heat from the lattice being used to decrease the spin ordering, leading to lattice cooling.

- Consider the paramagnetic system having N ions each of spin S (*i.e.*, with $2S+1$ states).
  - At temperature, T that is high enough to randomize the spins completely (overcoming any internal interactions, $k_B \Delta$ that may tend to orient the spins preferentially), the number of ways in which N spins can be arranged in $2S+1$ states is given by, $(2S+1)^N$. The spin entropy, $\sigma_S = k_B \ln (2S+1)^N = Nk_B \ln (2S+1)$. This spin entropy is reduced by the magnetic field which separates the $2S+1$ levels and populates the lower ones.

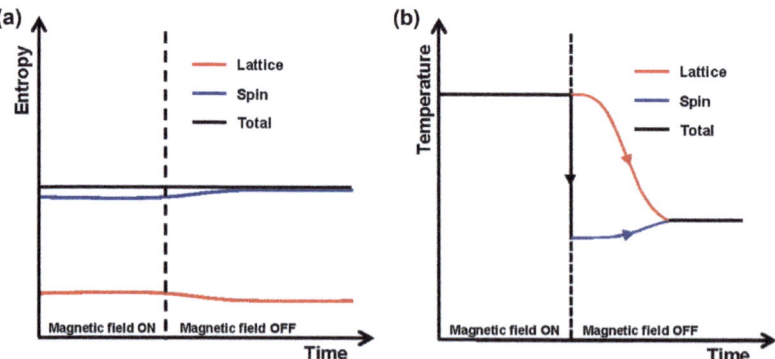

Figure 5.8 (a) Total entropy remaining constant during adiabatic demagnetization; lattice entropy decreases and spin entropy increases when the magnetic field is switched off under adiabatic condition. (b) Temperature decrease of the lattice overwhelms the temperature increase of the spin system upon adiabatic demagnetization leading to overall cooling. Drawn schematically based on Figure 7 in Chapter 14 of ref. 1.

○ Population of a magnetic sublevel is a function of $\dfrac{\mu H}{k_B T}$ (see Section 5.3.1), and hence $\dfrac{H}{T}$. As the spin entropy depends only on the population distribution, it is also a function of $\dfrac{H}{T}$. If the local interactions within the sample correspond to an internal field, H_Δ, the final temperature, T_2 that can be reached by adiabatic demagnetization, starting with an initial temperature T_1 is given by: $T_2 = T_1 \left(\dfrac{H_\Delta}{H} \right)$ where H is the external field.

○ As nuclear paramagnets have relatively lower magnetic moment $\left(\mu \propto \beta_N \propto \dfrac{1}{m} \right)$, the internal fields, H_Δ are smaller, allowing lower temperatures to be reached. Cu nuclei in the metal is a typical example for efficient adiabatic demagnetization.

5.4 Cooperative Phenomena – Ferro-, Antiferro-, and Ferrimagnetism

5.4.1 Exchange Energy

Exchange interaction is a quantum mechanical consequence of the identity of particles.

- Permutation symmetry: when two indistinguishable particles are interchanged, the sign of the wave function may change (antisymmetric) or remain unchanged (symmetric).
- Exchange interaction alters the expectation value of the repulsion energy between particles. For two fermions (overall wave function antisymmetric), the exchange interaction leads to a ground state with electron spins aligned parallel.
- Electrons are spin $\frac{1}{2}$ systems, and hence fermions. When electrons are exchanged with respect to space and spin coordinates, the total wave function changes sign.

Consider a 2-electron system (for example a H_2 molecule). With only the space coordinates, and the functions $\Phi_a(\vec{r}_1)$ and $\Phi_b(\vec{r}_2)$ in position space of electrons 1 and 2, the wave function of the system can be constructed as an antisymmetric combination, $\Psi_A(\vec{r}_1, \vec{r}_2)$ or symmetric combination, $\Psi_S(\vec{r}_1, \vec{r}_2)$. (Later, these are combined with the symmetric and antisymmetric spin wave functions respectively, to obtain the antisymmetric total wave functions.)

- $\Psi_A(\vec{r}_1, \vec{r}_2) = \dfrac{1}{\sqrt{2}} [\Phi_a(\vec{r}_1)\Phi_b(\vec{r}_2) - \Phi_a(\vec{r}_2)\Phi_b(\vec{r}_1)].$
- $\Psi_S(\vec{r}_1, \vec{r}_2) = \dfrac{1}{\sqrt{2}} [\Phi_a(\vec{r}_1)\Phi_b(\vec{r}_2) + \Phi_a(\vec{r}_2)\Phi_b(\vec{r}_1)].$

- The Hamiltonian for estimating the total energy is: $\mathcal{H} = \mathcal{H}^{(0)} + \mathcal{H}^{(1)}$; the second term can be considered as a perturbation on the atomic term represented by the first.
  - ○ Consider two electrons, singly occupied in two degenerate orbitals, Φ_a and Φ_b.
  - ○ When they are brought together from infinity to a distance r_{12} the repulsion energy is given by $\mathcal{H}^{(1)} = \dfrac{e^2}{r_{12}}$.

- The two eigen values for the system energy are calculated to be:
  - ○ $E_\pm = K_{ab} \pm J_{ab}$: E_+ and E_- are respectively the energy of the spatially symmetric and antisymmetric solutions. $K_{ab} =$ **Coulomb integral**, $J_{ab} =$ **exchange integral**.
  - ○ $K_{ab} = \displaystyle\int \left[\Phi_a^*(\vec{r}_1)\Phi_b^*(\vec{r}_2) \left(\dfrac{1}{r_{12}} \right) \Phi_a(\vec{r}_1)\Phi_b(\vec{r}_2) \right] d^3r_1 d^3r_2.$
  - ○ $J_{ab} = \displaystyle\int \left[\Phi_a^*(\vec{r}_1)\Phi_b^*(\vec{r}_2) \left(\dfrac{1}{r_{12}} \right) \Phi_a(\vec{r}_2)\Phi_b(\vec{r}_1) \right] d^3r_1 d^3r_2.$

- The symmetric and antisymmetric combinations of spin functions are:
  - ○ $\chi_S = [\alpha(1)\beta(2) + \alpha(2)\beta(1)]$
  - ○ $\chi_A = [\alpha(1)\beta(2) - \alpha(2)\beta(1)]$

- The energy eigen values E_+ and E_- correspond respectively to the overall wave functions, $\Psi_S(\vec{r}_1, \vec{r}_2) \cdot \chi_A$ (spatially symmetric/spin-singlet), and $\Psi_A(\vec{r}_1, \vec{r}_2) \cdot \chi_S$ (spatially antisymmetric/spin-triplet).

The **Heisenberg exchange Hamiltonian** represents the energy level splitting between E_+ and E_- as a spin Hamiltonian. For a system with spin sites i and j (or electrons in orbitals i and j), the exchange interaction is given in terms of the exchange energy, J:

- $E_{\text{Heisenberg}} = -J \langle \vec{S}_i \cdot \vec{S}_j \rangle$, where $\vec{S}_i$ and $\vec{S}_j$ are the spin operators.
- Note that the operator, $\vec{S}^2 = (\vec{S}_i + \vec{S}_j)^2 = \vec{S}_i^2 + 2\vec{S}_i \cdot \vec{S}_j + \vec{S}_i^2.$

- Therefore, $\vec{S}_i \cdot \vec{S}_j = \frac{1}{2}[\vec{S}^2 - (\vec{S}_i^2 + \vec{S}_i^2)]$.
- The corresponding eigen value is: $\frac{1}{2}[S(S+1) - (S_i(S_i+1) + S_j(S_j+1))]$.

For the two electrons in orbitals i, and j: $S_i = \frac{1}{2}$ and $S_j = \frac{1}{2}$; $S = 0$ (for the singlet state) and $S = 1$ (for the triplet state). Therefore, the eigen value for the $-J\langle \vec{S}_i \cdot \vec{S}_j \rangle$ operation is as follows.

- Singlet state: $-J\left\{\frac{1}{2}\left[0(0+1) - \left(\frac{1}{2}\left(\frac{1}{2}+1\right) + \frac{1}{2}\left(\frac{1}{2}+1\right)\right)\right]\right\}$
  $= -J \cdot \left(-\frac{3}{4}\right) = \frac{3}{4}J$.
- Triplet state: $-J\left\{\frac{1}{2}\left[1(1+1) - \left(\frac{1}{2}\left(\frac{1}{2}+1\right) + \frac{1}{2}\left(\frac{1}{2}+1\right)\right)\right]\right\}$
  $= -J \cdot \left(+\frac{1}{4}\right) = -\frac{1}{4}J$.
- Therefore, $E_{\text{Singlet}} - E_{\text{Triplet}} = J$. In the convention adopted for the equation for $E_{\text{Heisenberg}}$, if $J > 0$ the triplet is lower in energy than the singlet (Hund's rule).

Based on the earlier equations, the difference between the singlet and triplet state can be written as $E_+ - E_- = 2J_{ab}$.

- Combining with the treatment above for $E_{\text{Singlet}} - E_{\text{Triplet}}$, $J = 2J_{ab}$.

The exchange interaction can be visualized as providing an internal field, designated as the **exchange field**. This field plays the fundamental role in cooperative magnetic phenomena discussed in the following sections.

5.4.2 Ferromagnetism

Figure 5.9 shows a schematic plot (known as the Bethe–Slater curve) of how the exchange energy varies across some of the metallic elements, and is sensitively dependent on the ratio of the inter-atomic distance, a to the diameter of the 3d-shell, d. It clearly shows that only a fine balance of the structural and electronic features can lead to positive values of J.

In the previous section, it was shown that a positive J favours parallel orientation of electron spins (triplet state in the case of two electrons).

- Such an interaction tends to align the spins (and magnetic moments) of a collection of ions or atoms parallel, leading to **ferromagnetism**.

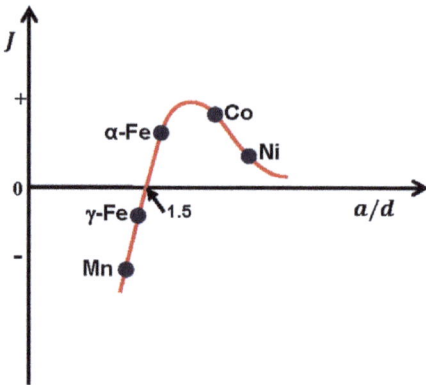

Figure 5.9   Schematic plot of exchange energy (*J*) *versus* the ratio of lattice
constant (*a*) to 3d-shell diameter (*d*) for different elements;
those with a positive value of *J* are ferromagnetic.

- Ferromagnets show a **spontaneous magnetization,** *i.e.*, magnetic moment in the absence of an applied magnetic field below a critical temperature. This is analogous to the case of ferro-electrics discussed in Section 4.5.2. The exchange interaction that leads to the parallel orientation of spins and gives rise to the spontaneous magnetization, can be visualized as an internal magnetic field, H_I.
- The orienting effect of H_I can be destroyed by increasing the temperature of the system above a threshold; the temperature at which the spontaneous magnetization vanishes is called the **Curie temperature,** T_C. The ordered ferromagnetic phase exists at $T < T_C$, and the disordered paramagnetic phase at $T > T_C$.
- The magnitude of the exchange field (internal field) can be as high as 10^7 Gauss.

   In the **mean field approximation**, each atom or ion experiences a magnetic field proportional to the magnetization of the system.

- The internal field is proportional to the magnetization of the system of atoms or ions: $H_I \propto M$, *i.e.*, $H_I = \lambda M$ where λ is a constant, independent of the temperature.
- In the paramagnetic phase $(T > T_C)$, assume that an external field, H_A is applied on the system. Magnetization now results from the applied as well as the internal field; the proportionality constant is a Curie (paramagnetic) susceptibility:

$$M = \chi_{para} \left(H_A + H_I \right).$$

- As the internal field is, in turn proportional to the magnetization,
  $$M = \chi_{\text{para}}(H_A + \lambda M) = \frac{C}{T}(H_A + \lambda M)\,(\text{using Curie law}).$$
- Since the magnetic susceptibility is the magnetic response of a system to an applied field, the equation can be rearranged to give: $\chi_{\text{para}} = \dfrac{M}{H_A} = \dfrac{C}{T - C\lambda}$. The equation is generally written as the **Curie–Weiss law**, $\chi_{\text{Curie–Weiss}} = \dfrac{C}{T - T_C}$, where $T_C = C\lambda$.
- The above equation shows that the magnetic susceptibility has a singularity at $T = T_C$. Diverging susceptibility means that at T_C, $M \neq 0$ even when $H_A = 0$, *i.e.*, magnetization exists in the absence of an applied field; this is spontaneous magnetization. This continues to grow as the temperature is lowered below T_C and saturates at $T = 0$ K (note that the above equation is derived for the paramagnetic phase, and is not applicable to the ferromagnetic phase).
- Plots of Curie–Weiss behavior (Figure 5.10(a)) shows χ *versus T* rising more sharply than the corresponding Curie curve (Figure 5.2(b)); the common method of plotting, $\dfrac{1}{\chi}$ *versus T* gives

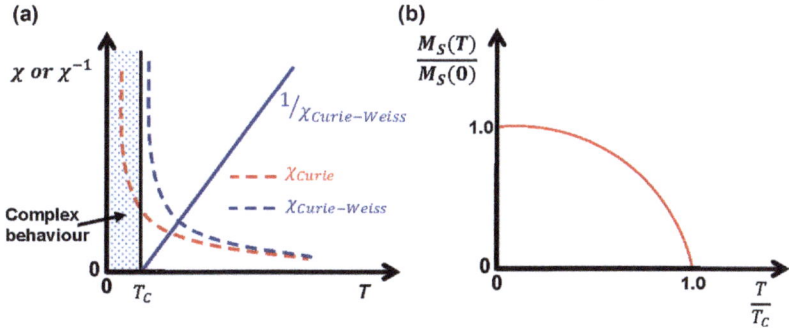

Figure 5.10 (a) Temperature dependence of magnetic susceptibility of paramagnets (Curie) compared to that of ferromagnets at $T > T_C$ (Curie–Weiss); the linear plot of inverse susceptibility of the latter is also shown. (b) Variation of saturation magnetization with temperature.

Table 5.2 Curie temperature of various ferromagnetic materials.

Material	Fe	Co	Ni	Gd	CrO$_2$	MnBi	Dy
T_C (K)	1043	1388	6272	292.5	386.5	630	88

in this case, a straight line with an x-intercept at T_C; below T_C, the susceptibility has a complex behavior.

The Curie temperatures of different ferromagnetic materials are shown in Table 5.2.

The maximum magnetization (M_s) due to the internal field (in the absence of an external field) at $T<T_C$ is called the **saturation magnetization.** M_s varies with temperature and vanishes at T_C (Figure 5.10(b)).

A qualitative band model provides insights into ferromagnetism. The model can be explained using the contrast between the neighboring elements, Ni (ferromagnet) and Cu (non-ferromagnetic, Pauli paramagnet).

- A Cu atom has the electronic configuration, [Ar] $3d^{10}4s^1$. The corresponding bands for Cu metal can be qualitatively shown as in Figure 5.11(a). The 3d band is filled with five electrons each of up and down spins, per atom. The half-filled 4s band has an equal number of up and down spin electrons. There is no spontaneous magnetization, and the metal is not ferromagnetic.
- An Ni atom has one electron less per atom, than Cu, *i.e.*, 10 electrons in the 3d and 4s bands together. The band structure at $T>T_C$ (Figure 5.11(b)) shows that the position of the Fermi surface is such that 0.54 electron are present in the 4s band and

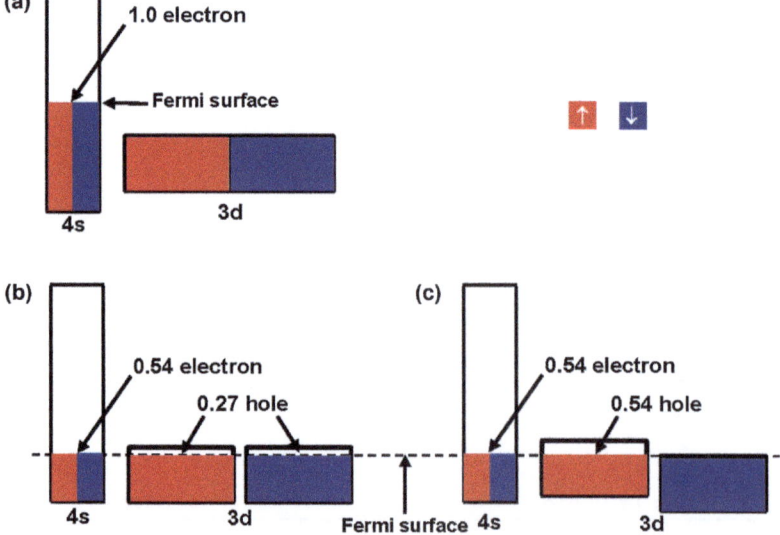

Figure 5.11   Schematic diagram of the 3d and 4s bands of (a) Cu metal, (b) Ni metal at $T>T_C$ and (c) Ni metal at 0 K. The up and down spin bands are indicated in red and blue respectively; the positions of Fermi surface and electron/hole count (per atom) are shown.

9.46 electrons in the 3d band (0.27 hole in each of the up and down spin bands).

- At 0 K, the exchange field present in the metal stabilizes one spin (say, down spin) over the other (up spin), so that the down spin band has five electrons per atom (fully filled) and the up spin band has 4.46 electrons (0.54 hole) per atom (Figure 5.11(c)). The excess of 0.54 down spin electrons per atom leads to the spontaneous magnetization and ferromagnetic character. These electron-hole counts are arrived at from a measure of the spontaneous magnetization of Ni.

5.4.3 Hysteresis

At temperatures below T_C, the electronic magnetic moments of a ferromagnetic material are mostly aligned within microscopic regions called **domains**. Within the domains, the magnetic moment would be close to the saturation value expected for the number of spin sites and the corresponding spin moments; these are regions of spontaneous magnetization. The origin of the domains is discussed in Section 5.4.4.

In the sample as a whole (at the macroscopic scale) however, the magnetization may be much less than the saturation value, or even zero. This means that the magnetization of the various domains cancel each other. The magnetization, M of a ferromagnetic material at $T < T_C$ varies as shown in Figure 5.12, when the sample is subjected to an external magnetic field, H. A similar curve can also be plotted for

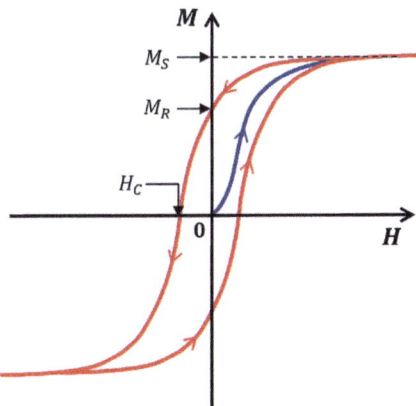

Figure 5.12 Magnetization (M) *versus* magnetic field (H) curve for a ferro-magnet at temperatures below the Curie temperature, T_C. The saturation magnetization (M_S), remanence (M_R) and coercive field (H_C) are indicated. The initial magnetization (blue curve) and the hysteresis cycle (red curve) are shown.

magnetic induction, *B versus* field; recall also the case of ferroelectrics in Figure 4.34.

- The magnetization increases when the sample is subjected to an increasing magnetic field, steeply first, and more gradually later. It reaches the saturation value, M_S at sufficiently high applied fields; thereafter, any increase in the field does not change the magnetization.
- When the field is now reduced, the magnetization decreases gradually, but does not return along the same path as when the field was increased. As the field goes to zero, the magnetization remains non-zero; this value is called the remanence, M_R.
- Now, as the field is increased in the opposite direction, the magnetization decreases steadily until it reaches zero; the field at which the magnetization goes to zero is known as the coercive field, H_C.
- As the field in the opposite direction is increased further, the magnetization increases as before, but now in the opposite direction, reaching the saturation value. The process can be repeated, giving rise to the **hysteresis cycle**.
- The area enclosed by the *M–H* curve, is the energy dissipated during the magnetization–demagnetization cycle.

The origin of the hysteresis cycle can be traced to the growth and reorientation of the magnetic domains in the ferromagnetic sample.

5.4.4 Magnetic Domains

It was mentioned in the previous section that, even though one would expect all the spin moments in a ferromagnetic sample at $T < T_C$ to be aligned parallel (owing to the exchange interactions), the sample may show nearly zero bulk magnetization, unless deliberately magnetized. This was attributed to the domain structure and the sum of the magnetization of all the domains giving a net zero value.

The domain structure results from a balance between various energies, primarily, exchange and magnetic dipolar; magnetocrystalline (anisotropy) and magnetostriction are other factors that influence the domain formation.

- The exchange interaction is strong, tending to align adjacent spins parallel. However, they are short-range, their influence falling off exponentially with distance ($\propto e^{-r}$).

- On the other hand, dipolar interactions (between the magnetic dipoles, tending to align them antiparallel), persist over longer distances, decreasing as the inverse cube of the distance $\left(\propto \dfrac{1}{r^3}\right)$.
- Minimum total magnetic energy is achieved by the formation of magnetic domains; each domain possesses saturation magnetization by orienting all spins parallel within (the consequence of short-range exchange interaction), but the magnetic moments of the collection of domains are oriented randomly, so as to have no net magnetic moment for the specimen (consequence of long-range dipolar interaction).
- The spin orientations across the boundary between two domains with opposite spin moments (Figure 5.13) illustrate how the short-range exchange interaction between near neighbors (spins oriented parallel), as well as long-range dipolar interactions between domains (magnetic moments oriented antiparallel) are satisfied. The region across which the spin orientation changes is known as the **Bloch wall**; in iron, it is ~0.1 μm (1000 Å) thick.
- Magnetocrystalline or anisotropy energy results from the ease of magnetization along selected crystal axes. Magnetostriction refers to the elongation or shrinkage of ferromagnetic samples upon magnetization. These factors can influence the domain formation, as the spin orientations vary across the domain boundaries.

The hysteresis shown in Figure 5.12 can be understood in terms of the behavior of the domains under the influence of an external magnetic field.

- Figure 5.14(a) is a schematic representation of the domain structure that results in zero magnetic moment in an unmagnetized ferromagnetic sample.
- Figure 5.14(b) and (c) shows how application of an increasing external field leads to growth of one domain at the expense of the others, leading eventually to saturation of the magnetization of the sample; the domain magnetization rotates at higher magnetic fields to align with the direction of the applied field (Figure 5.14(d)).

Domain 1   **Bloch wall**   **Domain 2**

Figure 5.13   Schematic representation of the spin orientation in two magnetic domains and their variation across the boundary labeled as Bloch wall.

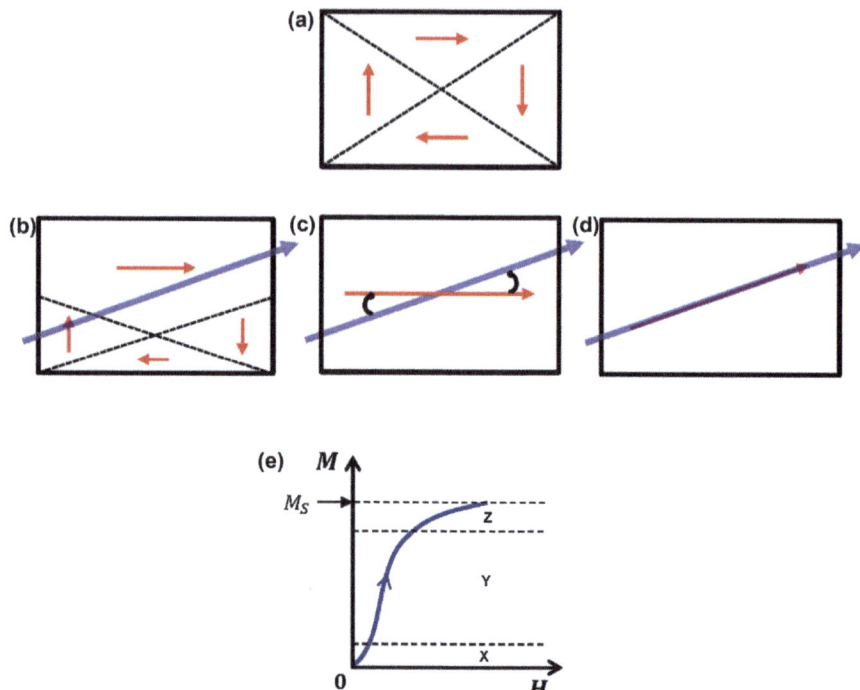

Figure 5.14 Schematic diagram showing the domain structure of a ferromagnetic specimen (red arrow: magnetic moment of each domain; blue arrow: applied external magnetic field; broken line: domain boundary wall): (a) unmagnetized state, (b) growth of one domain at the expense of others in an applied magnetic field, (c) saturation of magnetization, (d) magnetization reorientation. (e) Magnetization *versus* applied magnetic field; the different regimes are indicated as, X: reversible boundary displacement, Y: irreversible boundary displacement, Z: magnetization rotation. Drawn schematically based on Figure 7.20 and 7.21 in ref. 2.

- The associated magnetization change with the field is shown in Figure 5.14(e).
- The above steps constitute the process of **magnetization** of a ferromagnetic sample.
- When the applied magnetic field is reduced to zero, the system does not go back to the original domain structure with zero magnetization. In other words, the magnetization curve is not retraced back; some magnetization, M_R remains. The reason for this is that, the formation of the domain boundaries requires energy. Increasing the magnetic field in the opposite direction up to the coercive field, H_C, provides this activation, leading to zero net magnetization.

Permanent magnets require materials with a high coercive field; Fe–Pt is a typical example, with $H_C \sim 120$ kG. Transformer cores on the other hand, require magnets with high permeability and low coercive fields, so that hysteresis losses during magnetization–demagnetization cycles are low; a typical example is Si–Fe with $H_C \sim 0.5$ G.

5.4.5 Antiferromagnetism

In the discussion of the Heisenberg Hamiltonian, $E_{\text{Heisenberg}} = -J \langle \vec{S}_i . \vec{S}_j \rangle$ in Section 5.4.1, it was shown that $E_{\text{Singlet}} - E_{\text{Triplet}} = J$. When the exchange energy, J is positive, a ferromagnetic coupling of spins occurs. If J is negative, it would imply that the singlet coupling of the spins is energetically favored over the triplet coupling. Such spin interactions in a bulk material lead to antiparallel coupling of neighboring spins, giving rise to antiferromagnetism.

The spin organization in an antiferromagnet can be illustrated using the example of MnO, which has a NaCl crystal structure, with interpenetrating fcc lattices of Mn^{2+} and O^{2-} ions.

- Figure 5.15(a) shows the Mn^{2+} ion lattice; alternating layers of Mn^{2+} have opposite spins. It can be visualized as having sub-lattices A and B, with opposite spin orientations.

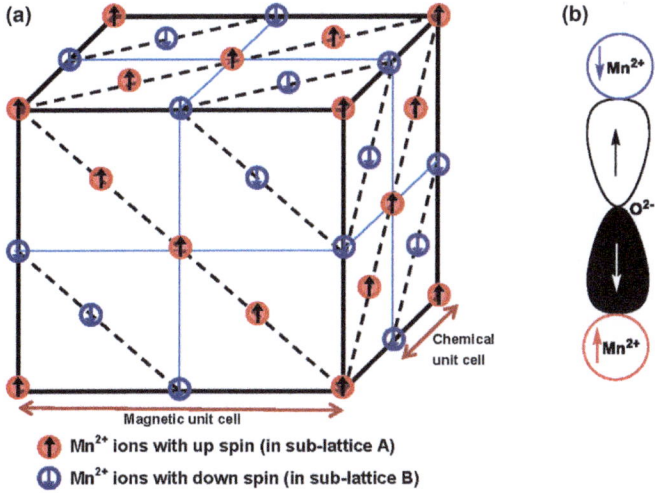

Magnetic unit cell

⬆ Mn^{2+} ions with up spin (in sub-lattice A)

⊕ Mn^{2+} ions with down spin (in sub-lattice B)

Figure 5.15 (a) Lattice of Mn^{2+} ions with their spin orientation in the antiferromagnetically ordered state of MnO. The sub-lattices are indicated; oxide ions are omitted for clarity. (b) Schematic figure showing the superexchange between two Mn^{2+} ions (one from each sub-lattice, typically involving the d-orbitals), mediated by the O^{2-} ligand orbital, leading to the antiferromagnetic coupling of their spins.

- The spin lattice can be observed using neutron scattering (Section 1.5.1); in the antiferromagnetically ordered state, a unit cell length that is double the chemical unit cell (observed in X-ray scattering experiment) is observed. This situation arises since the neutrons (having magnetic moment) distinguish the spin orientation of the ions and treat those with opposite spins as different species, while the X-ray photons do not distinguish the spin orientations.
- The net magnetization in the ordered state is zero.
- The antiferromagnetic spin coupling can be explained using the '**superexchange interaction**' between the spins of the metal ions (Mn^{2+}) mediated by the electrons in the filled orbital of the ligand ion (O^{2-}), shown schematically in Figure 5.15(b).

As in the case of ferromagnetic materials (Section 5.4.2), antiferromagnetic materials show the ordered state only at temperatures that are low enough where the thermal fluctuations do not overcome the exchange interactions. The temperature below which the ordered antiferromagnetic state is formed is called the **Néel temperature**, T_N.

- When $T > T_N$, the spin orientations are random; the resulting paramagnetic state together with an internal field that tends to orient neighboring spins antiparallel, can be described by extending the mean field approach developed in Section 5.4.2 to a two sublattice model.
- Now, $H_I \propto -M$, so that the final expression for the paramagnetic susceptibility will be, $\chi_{para} = \dfrac{C}{T + C\lambda}$. The corresponding Curie–Weiss law for antiferromagnetic materials at $T > T_N$ is usually written as: $\chi_{Curie-Weiss} = \dfrac{C}{T + \theta}$, where $\theta = C\lambda$.
- The variation of magnetic susceptibility with temperature for an antiferromagnet is shown in Figure 5.16. At $T > T_N$, the Curie–Weiss law is obeyed; the inverse susceptibility plot shows a negative x-intercept.
- When cooled below T_N, the material undergoes a phase transition to the ordered antiferromagnetic state. The susceptibility goes to zero at 0 K, if the magnetic field is parallel to the spin axis, but decreases to a finite value if the field is perpendicular.

5.4.6 Ferrimagnetism

If the spin of the metal ions in the two sub-lattices A and B are not equal (or if the two sub-lattices have identical but unequal number of

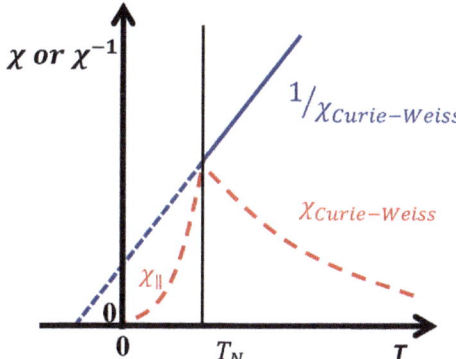

Figure 5.16 Temperature dependence of magnetic susceptibility of antifer-romagnets. Curie–Weiss law is followed at $T > T_N$; the linear plot of inverse susceptibility is also shown (blue line). The suscepti-bility component for the magnetic field parallel to the spin axis decreases to zero at $T < T_N$; the perpendicular component de-creases to a finite value (not shown).

magnetic ions per unit cell), and there exists antiferromagnetic coupling between them, the material will show a net magnetization in the ordered state. Such materials are called ferrimagnetic. Classic examples of ferrimagnets are ferrites with the general formula, $MO \cdot Fe_2O_3$, where $M = Zn$, Cd, Fe, Ni, Cu, Co, or Mg in $+2$ oxidation state. Magnetite, Fe_3O_4 (*i.e.*, $FeO \cdot Fe_2O_3$) is the most famous example.

- In Fe_3O_4, one Fe^{3+} occupies tetrahedral (T_d) sites whereas the second Fe^{3+} and the Fe^{2+} ions are in octahedral (O_h) sites formed by the fcc (or ccp) lattice of O^{2-} ions. The formula can be written as $(Fe^{3+})^t(Fe^{3+}, Fe^{2+})^o(O^{2-})_4^{ccp}$. In the spinel structure, AB_2O_4, oxide ions form an fcc lattice and A^{2+} occupy $\frac{1}{8}$ of the T_d sites and B^{3+} occupy $\frac{1}{2}$ of the O_h sites; Fe_3O_4 has an inverse spinel structure, A^{2+} occupying O_h sites and B^{3+} occupying T_d and O_h sites.
- For Fe^{2+}, the spin, $S = 2$, whereas for Fe^{3+}, $S = 5/2$. The formula Fe_3O_4 alone would suggest a magnetic moment of $\left\{\left(2 \times \frac{5}{2}\right) + 2\right\}g\beta = 7g\beta$. However, the antiferromagnetic coup-ling between the ions in the two sub-lattices, T_d and O_h sites leads to a net magnetic moment of $2g\beta$ (Figure 5.17(a)).
- The inverse magnetic susceptibility behaves in a complex manner (Figure 5.17(b)), appearing antiferromagnetic at higher

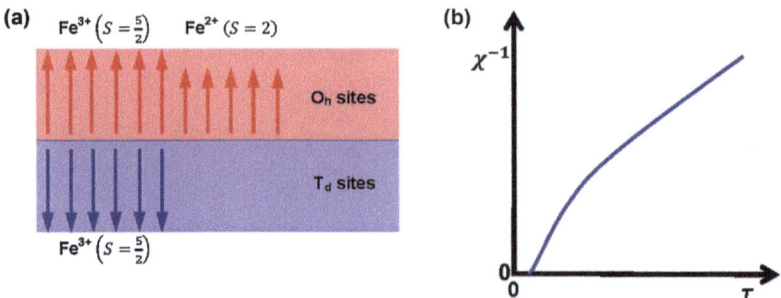

Figure 5.17 (a) Schematic diagram showing the sub-lattices with unequal magnetic moments in the ferrimagnet, Fe_3O_4. (b) Temperature dependence of magnetic susceptibility of ferrimagnets.

temperature (Figure 5.16), but at low temperatures, reaching a positive intercept as in a ferromagnet (Figure 5.10(a)).

- Iron garnets, $M_3Fe_5O_{12}$ (M is a trivalent ion) are also ferrimagnets; yttrium–iron garnet, $Y_3Fe_5O_{12}$ is well-known. Y^{3+} is diamagnetic; Fe^{3+} $(S = \dfrac{5}{2}$, moment $= \dfrac{5}{2}g\beta)$ are distributed, three in T_d and two in O_h sites. The antiferromagnetic coupling between the two sub-lattices results in a net magnetic moment of $\dfrac{5}{2}g\beta$.

5.4.7 Superparamagnetism

Superparamagnetism is a type of magnetism that appears in nano-particles of ferromagnetic or ferrimagnetic materials.

Particles of ferro/ferrimagnetic materials with sizes in the range of a few nanometers to a few tens of nanometers (nanoparticles) exist as single domain systems.

- Magnetization of the individual particles corresponds to a single giant magnetic moment, the sum of the magnetic moments carried by constituent atoms/ions.
- The magnetization can take alternate directions with respect to the easy axis determined by the magnetic anisotropy (Section 5.4.4). It can randomly flip under the influence of tem-perature (when the energy due to magnetocrystalline anisotropy is less than the thermal energy). The time between flips is known as the **Néel relaxation time**, τ_N.
- If magnetic measurements are made over time periods signifi-cantly longer than τ_N, the magnetization of the particles averages to zero; the system is said to be in the superparamagnetic state.

All ferro/ferrimagnetic materials exist in a paramagnetic state above their Curie temperature (Sections 5.4.2 and 5.4.6); a superparamagnetic state exists below the Curie temperature, and is therefore distinctly different.

- The magnetization measured in an external magnetic field and hence the magnetic susceptibility of the superparamagnetic state is significantly higher than that in the normal paramagnetic state above the T_C.
- The temperature below which, the magnetization relaxation time equals the time scale of the experimental technique employed, is called the **blocking temperature**.

5.5 Experimental Determination of Magnetic Susceptibility

Measurement of magnetic susceptibility, often at different temperatures, is the most fundamental characterization of magnetic materials. As discussed in the earlier sections, the temperature dependence of magnetic susceptibility determines the general type of magnetism present in the material.

SQUID (Section 4.4.5) provides one of the most sensitive and accurate methods to measure magnetic fields; SQUID magnetometers are commonly used in contemporary magnetic susceptibility measurements. However, the classic methods for the measurement of magnetic susceptibility are based on the Gouy and Faraday balances. The general principle is based on measuring the force on a magnetic material in an external field.

- Magnetization can be defined based on the response of the free energy of the system in the magnetic field (see Section 5.6),

$$M = -\frac{1}{V}\left(\frac{\partial G}{\partial \boldsymbol{H}}\right)_T.$$

- For a specimen in an inhomogeneous magnetic field (H varying with position x), the change in free energy can be written as:

$$\mathrm{d}G = G[H(x+\mathrm{d}x)] - G[H(x)] = \left(\frac{\partial G}{\partial H}\right)\left(\frac{\partial H}{\partial x}\right)\mathrm{d}x = -VM\left(\frac{\partial H}{\partial x}\right)\mathrm{d}x$$

$$\therefore \frac{\partial G}{\partial x} = -VM\left(\frac{\partial H}{\partial x}\right).$$

- The force acting on the specimen can be related to the work done and hence the free energy change as, $F = -\dfrac{dG}{dx}$. Therefore, $\dfrac{F}{V} = -\dfrac{1}{V}\left(\dfrac{dG}{dx}\right)$.

- Incremental force, over a small volume change can be written as:

$$\frac{dF}{dV} = -\frac{1}{V}\left(\frac{dG}{dx}\right) = M\left(\frac{\partial H}{\partial x}\right) \Rightarrow dF = \chi H\left(\frac{\partial H}{\partial x}\right)dV,$$

where χ is the volume susceptibility.

5.5.1 Gouy Method

The sample taken in a long tube of length, l (in order to ensure sensitivity, large samples are required) is inserted inside a magnetic field and the force in terms of the change in weight, mg (Figure 5.18) is measured. The magnetic field varies along the length of the tube (pole pieces are designed so that $\dfrac{\partial H}{\partial x}$ is constant).

- The expression derived above can be rewritten in terms of the cross-sectional area, A of the sample: $dF = \chi H\left(\dfrac{\partial H}{\partial x}\right)(A \cdot dx)$ $= \chi A \cdot H dH$.

- Integration over the varying field on the sample gives an expression for the force: $F = \chi A \displaystyle\int_{H_1}^{H_2} H dH = \dfrac{\chi A}{2}\left(H_2^2 - H_1^2\right)$

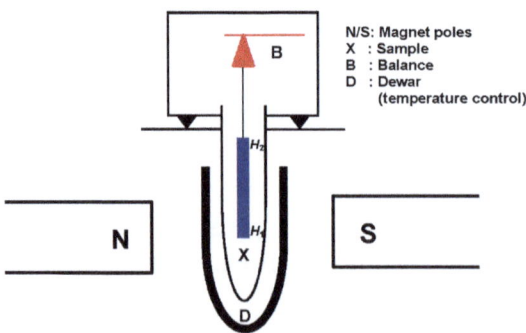

N/S: Magnet poles
X : Sample
B : Balance
D : Dewar
    (temperature control)

Figure 5.18 Schematic diagram of a Gouy balance for magnetic suscepti-bility measurement. The magnetic field variation (from H_1 to H_2) along the sample is indicated.

$$= \frac{\chi_g \rho V}{2l}\left(H_2^2 - H_1^2\right) = \frac{\chi_g W}{2l}\left(H_2^2 - H_1^2\right), \text{ where } \chi_g \text{ is the gram sus-}$$
ceptibility; ρ, W are the density and weight of the sample.

- Rewriting the force in terms of the change in weight, $mg = \frac{\chi_g W}{2l}\left(H_2^2 - H_1^2\right)$, the gram susceptibility can be written as:

$$\chi_g = \frac{2mgl}{W\left(H_2^2 - H_1^2\right)} = \frac{C'm}{W} \text{ where } C' \text{ is a constant depending on the}$$
sample length and magnetic field applied.

- If the weight changes measured are m_R and m_S for a reference sample, R (with known magnetic susceptibility, χ_g^R) and specimen, S with weights W_R and W_S respectively, the susceptibility of the specimen can be determined:

$$\frac{\chi_g^S}{\chi_g^R} = \left(\frac{C'm_S}{W_S}\right) \Big/ \left(\frac{C'm_R}{W_R}\right) \Rightarrow \chi_g^S = \left(\frac{m_S}{m_R}\right)\left(\frac{W_R}{W_S}\right)\chi_g^R$$

5.5.2 Faraday Method

In a Faraday balance, the magnetic poles are designed in such a way that the product of the field strength and field gradient in the x-direction, $H\left(\dfrac{\partial H}{\partial x}\right)$ is constant (Figure 5.19). Small sample sizes are used, and the effect of the field over the sample is homogeneous.

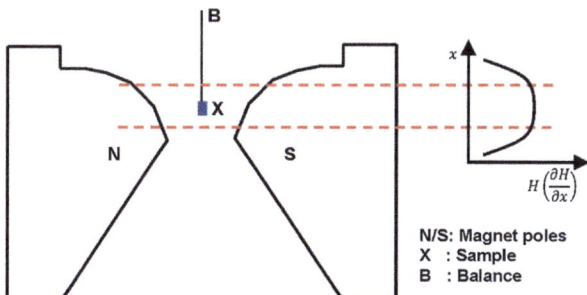

Figure 5.19 Schematic diagram of a Faraday balance for magnetic susceptibility measurement. The specially constructed pole pieces are shown; the Dewar for temperature control is omitted for clarity. The plot shows the variation of the product of the field strength and field gradient in the x-direction; the sample X resides in the constant value regime.

- Under this situation, the earlier equation can now be written as:

$$\frac{F}{V} = \chi H \left(\frac{\partial H}{\partial x} \right) = C\chi = C\chi_g \frac{W}{V}.$$

- Therefore, $\chi_g = \dfrac{F}{CW} = \dfrac{mg}{CW}$.
- As described above, the value of C can be determined by calibration using a reference sample of known susceptibility.

5.5.3 Vibrating Sample Magnetometer

Magnetic susceptibility can be measured using an approach different from the measurement of force using the balances discussed above. This is based on Faraday's law of electromagnetic induction, generation of an electromotive force (E) in a wire loop when the magnetic flux (ϕ_B) through the surface enclosed by the loop varies with time,

$$E = -\frac{d\phi_B}{dt}.$$

In a vibrating sample magnetometer, the sample is magnetized using a magnetic field; the extent of magnetization will depend on the susceptibility of the material. When the sample is then vibrated sinusoidally, resulting variation of the magnetic flux induces an EMF in the pickup coils around it. The electric field produced in the coils provides a measure of the magnetization of the sample and hence its susceptibility. Variation of the magnetization of the sample upon changing the applied magnetic field and temperature, provides information on the magnetic characteristics of the sample.

5.6 Basic Definitions

Some basic definitions relevant to the discussions in this chapter are listed below.

- **Magnetic field** is a vector field (a vector assigned to each point in space) that describes the magnetic influence from external electric currents in a material, independent of the material's magnetic response. H has the unit of amperes per meter (A m^{-1}) and describes the force acting on a unit magnetic moment; the older unit was the oersted (1 A m^{-1} = 0.01257 oersted).
- **Magnetization, M** of a material is the magnetic moment induced per unit volume of the material by an applied field, H. Magnetic

susceptibility (or simply susceptibility) is defined as, $\chi = \dfrac{\partial \mathbf{M}}{\partial \mathbf{H}}$. A linear relationship between $\mathbf{M}$ and $\mathbf{H}$ implies $\mathbf{M} = \chi \mathbf{H}$. Magnetization can be defined in terms of the free energy G of a material in a magnetic field at temperature T as $\mathbf{M} = -\dfrac{1}{V}\left(\dfrac{\partial G}{\partial \mathbf{H}}\right)_T$; the susceptibility can be written as $\chi = -\dfrac{1}{V}\left(\dfrac{\partial^2 G}{\partial \mathbf{H}^2}\right)_T$.

- **Magnetic induction** or magnetic flux density is defined as $\mathbf{B} = \mu_0(\mathbf{H} + \mathbf{M})$, where μ_0 is called the vacuum permeability. This equation is in SI units; in CGS units, it would be: $\mathbf{B} = \mu_0(\mathbf{H} + 4\pi\mathbf{M})$. $\mathbf{B}$ represents the field due to external currents, as well as that generated in a material. The unit of $\mathbf{B}$ is gauss or tesla ($1\ \mathrm{T} = 10^4\ \mathrm{G}$). The force exerted on a particle with charge q moving with velocity v in the magnetic field is given by $\mathbf{F} = q(\mathbf{v} \times \mathbf{B})$ in SI units ($\mathbf{F} = \dfrac{q}{c}(\mathbf{v} \times \mathbf{B})$ in CGS units, c = velocity of light).

- A relation commonly used is $\mathbf{B} = \mu_m\,\mathbf{H}$ (Note: $\mathbf{B} = \mu_0\,\mathbf{H}$ in a vacuum). Since $\mathbf{B} = \mu_0(\mathbf{H} + \mathbf{M}) = \mu_0\,(1 + \chi)\,\mathbf{H}$, $\mu_m = \mu_0(1 + \chi) = \mu_0 K_m$, where K_m is known as the relative permeability.

- Since $\mathbf{M}$ is defined per unit volume, χ can be referred to as volume susceptibility. Gram susceptibility, $\chi_g = \dfrac{\chi}{\rho}$, where ρ is the density of the material, and molar susceptibility, $\chi_m = \chi_g \times \mathrm{MW}$, MW is the molecular weight of the material.

References

1. C. Kittel, *Introduction to Solid State Phyiscs*, 5th Edition, Wiley, New York, 1976.
2. G. I. Epifanov, *Solid State Physics*, Mir Publishers, Moscow, 1979.

6 Optical Properties

6.1 Optical Reflectance, Absorption and Scattering

Optical properties are some of the most important materials attributes, as they include the first visual characterization of the material. Color and opacity of materials are determined by their reflectivity and absorptivity at different frequencies of visible light. Part of the incident radiation that is neither reflected nor absorbed accounts for the transmittivity of the material. It is important to study the different forms of interaction of photons with materials, involving reflection or transmission of light, absorption, emission, scattering, and even ionization when the photons have sufficiently high energy or intensity.

6.1.1 Reflectivity

Reflectivity can be defined as, $R(\omega) = \dfrac{I_r}{I_i} = \dfrac{|E_r|^2}{|E_i|^2}$, where I_i and I_r are the intensity of incident and reflected light at the frequency, ω and E_i and E_r are the amplitudes of the corresponding electric fields.

Absorptivity and transmittivity can similarly be defined as, $A(\omega) = \dfrac{I_a}{I_i}$, and $T(\omega) = \dfrac{I_t}{I_i}$ respectively; I_a and I_t are the intensity of light absorbed and transmitted respectively. Since, $I_i = I_r + I_a + I_t$, $R(\omega) + A(\omega) + T(\omega) = 1$.

- The complex dielectric function at frequency, ω is defined as, $\kappa(\omega) = n(\omega) + ik(\omega)$; $n(\omega)$, the real part is the refractive index

Core Concepts for a Course on Materials Chemistry
By T. P. Radhakrishnan
© T. P. Radhakrishnan 2023
Published by the Royal Society of Chemistry, www.rsc.org

and $k(\omega)$, the imaginary part is the extinction coefficient. $\kappa(\omega) = \sqrt{\epsilon(\omega)}$ where $\epsilon(\omega)$ is the dielectric constant.

- The physical meaning of the complex dielectric function can be seen in the definition, $\kappa(\omega) = \dfrac{E_i - E_r}{E_i + E_r}$, where the numerator represents the electric field inside the material and the denominator, the electric field outside.

  ○ From the above definition, (frequency is not stated explicitly in the following equations),

  $$(E_i + E_r)\kappa = E_i - E_r \Rightarrow (1+\kappa)E_r = (1-\kappa)\,E_i \Rightarrow \frac{E_r}{E_i} = \frac{(1-\kappa)}{(1+\kappa)}$$

  $$\therefore R = \frac{|E_r|^2}{|E_i|^2} = \frac{|1-\kappa|^2}{|1+\kappa|^2} = \frac{|1-(n+ik)|^2}{|1+(n+ik)|^2} = \frac{(1-n-ik)(1-n+ik)}{(1+n+ik)(1+n-ik)}$$

  {Note: $|x|^2 = x.x^*$}

  ○ This leads to (including frequency explicitly): $R(\omega) = \dfrac{(1-n(\omega))^2 + (k(\omega))^2}{(1+n(\omega))^2 + (k(\omega))^2}$, showing that the reflectance at frequency ω depends on the refractive index and extinction coefficient at that frequency.

6.1.2 Dielectric Constant

When a medium with N_i charge carriers, q_i is placed in an electric field, E, the displacements of the carriers, x_i lead to the dielectric polarization (net charge displacement per unit volume), $P = \sum\limits_i N_i q_i x_i$.

The phenomenon has been noted in connection with dielectric materials earlier (Section 4. 5).

At frequency, ω, the polarization can be related to the electric field as, $P(\omega) = \chi(\omega)E(\omega)$ where $\chi(\omega)$ is the electric susceptibility. The linear relation applies at relatively low fields; at high fields, the general relation is: $P(\omega) = \sum \chi_n(\omega)[E(\omega)]^n$ where $\chi_n(\omega)$ are the nonlinear susceptibilities of various orders (Section 6.6).

The displacement $D(\omega)$ is defined as: $D(\omega) = E(\omega) + 4\pi P(\omega) = \varepsilon(\omega)E(\omega)$ where ε is the dielectric constant. Substituting the linear relation for $P(\omega)$ shows that, $\varepsilon(\omega) = 1 + 4\pi\chi(\omega)$. For a free electron gas, an expression for $\varepsilon(\omega)$ can be derived as follows.

- The electrons are accelerated in an electric field; equating the product of mass and acceleration to the Lorentz force (Section 4.1.5),

  $$m\frac{\partial^2 x}{\partial t^2} = -eE$$

- The displacement and field can be written as: $x = x_0 e^{-i\omega t}$ and $E = E_0 e^{-i\omega t}$. Substituting in the above equation, $m \dfrac{\partial^2 (x_0 e^{-i\omega t})}{\partial t^2} = -e E_0 e^{-i\omega t}$.

- $\therefore\ m\omega^2 x = eE \Rightarrow x = \dfrac{eE}{m\omega^2}$

- The polarization due to N electrons with charge, e is given by, $P = -Nex = \dfrac{-Ne^2 E}{m\omega^2}$

- This leads to, $\chi(\omega) = \dfrac{-Ne^2}{m\omega^2}$. Therefore dielectric constant, $\varepsilon(\omega) = 1 - \dfrac{4\pi Ne^2}{m\omega^2}$.

- Defining plasma frequency, $\omega_p = \left(\dfrac{4\pi Ne^2}{m}\right)^{1/2}$, one can write, $\varepsilon(\omega) = 1 - \dfrac{\omega_p^2}{\omega^2}$.

- The plot of the function (Figure 6.1) shows that an electromagnetic wave with $\omega < \omega_P$ is absorbed in the medium, but when $\omega > \omega_P$, the wave is transmitted.

- This can be understood based on the equation, $\sqrt{\epsilon(\omega)} = n(\omega) + ik(\omega)$. For $\omega < \omega_P$, $\varepsilon(\omega) < 0$ and hence its square root is imaginary; the wave is absorbed (extinction coefficient, $k(\omega)$). For $\omega > \omega_P$, $\varepsilon(\omega) > 0$ and hence its square root is real; the wave is transmitted (refractive index, $n(\omega)$).

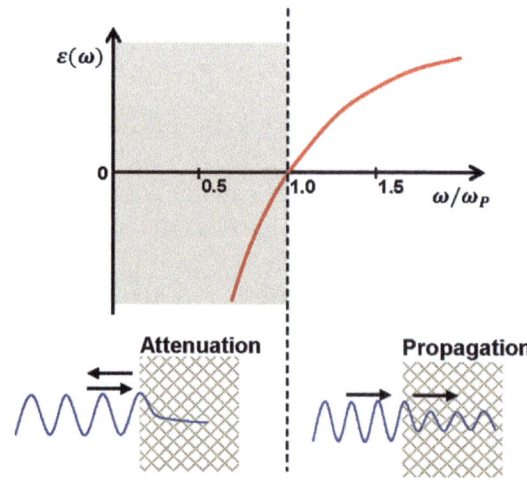

Figure 6.1 Schematic plot of the dielectric function of a free electron gas for electromagnetic waves of frequency (ω), in units of the plasma frequency (ω_P); reflection of the waves from the material, for $\omega < \omega_P$ and hence negative $\varepsilon(\omega)$, and its propagation for $\omega > \omega_P$ and positive $\varepsilon(\omega)$ are indicated. Drawn schematically based on Figure 1 in Chapter 10 in ref. 1.

For most metals, the plasma frequency is in the UV region; for example, for bulk gold metal, $v_P \sim 2\times10^{15}$ s^{-1} ($\omega_P \sim 1.3\times10^{16}$ s^{-1}), *i.e.*, the wavelength, $\lambda \sim 150$ nm.

- Light with a wavelength longer than this value ($\omega < \omega_P$) is reflected and attenuated inside the medium and light with a shorter wavelength ($\omega > \omega_P$) is transmitted.
- The collective oscillation of conduction electrons is called **plasmon** and the frequency of the oscillation is the **plasmon frequency**; the internal field set up by the oscillation of the electrons cancels the field due to the electromagnetic radiation.
- The shielding is complete for $\omega < \omega_P$, and the field is attenuated; for $\omega > \omega_P$, the frequency of the electromagnetic wave is too high for the collective electron oscillation to respond, and the wave is transmitted.

In spherical particles, as the radius (r) decreases, the ratio of surface area (A) to volume (V) increases $\left(\dfrac{A}{V} = \dfrac{4\pi r^2}{\dfrac{4}{3}\pi r^3} = \dfrac{3}{r} \right)$. Many nanoparticles (with a high A/V ratio as the radii are in the nanometer range) exhibit prominent responses due to surface plasmons. Strong absorption of electromagnetic radiation with energy matching the surface plasmon frequency is known as **surface plasmon resonance absorption.**

6.1.3 Excitons

Absorption of photons by semiconductors gives useful information about the band structure; the absorption occurs at all k points in the Brillouin zone (Section 4.2.2) for which photon energy, $\hbar\omega = E_c(k) - E_v(k)$, the electronic transition occurring from the valence to the conduction band. Optical absorption of semiconductors was discussed in Section 4.3.2. Absorption of photons leads to the formation electrons and holes. A bound electron–hole pair is known as an **exciton.**

- The binding energy results from the electrostatic interaction between the electron and the hole. So it depends on the spatial separation of the pair; it ranges from 1 meV to 1 eV (the latter in organic semiconductors).
- The exciton shows up as a structure in the absorption or reflectance spectra at energies below the band gap (Figure 6.2).
- **Frenkel excitons** are tightly bound electron–hole pairs, the electron and hole existing together on the same atom or molecule.

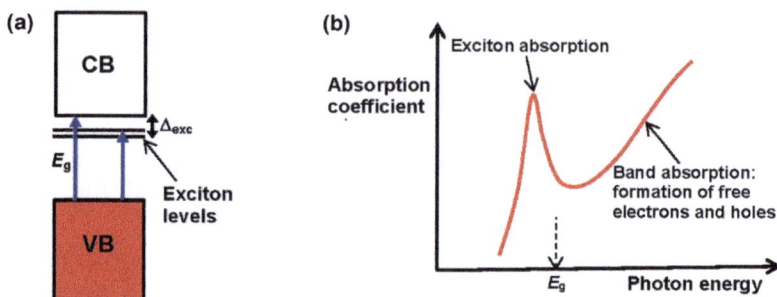

Figure 6.2 (a) Exciton energy levels in relation to the valence and conduction bands (VB, CB), indicating the energy gap, E_g and the exciton binding energy, Δ_{exc}. (b) A schematic optical absorption spectrum of a semiconductor showing the exciton absorption at energies just below the band absorption.

They are essentially the excited state of the atom or molecule, that can hop from one site to another. Frenkel excitons are found in crystals of inert gases as well as molecular crystals. In the latter, the excitation of single molecules appear as excitons in the crystals; a standard example is anthracene crystal with low-lying exciton states at 1.80 and 3.15 eV.

- **Wannier excitons** are weakly bound excitons. The electron and hole in the bound pair are significantly separated in the crystal lattice. They are found in semiconductors like Cu_2O.

In aggregates of molecules and molecular crystals, coupling of excitons can lead to new states with energies higher and lower compared to the reference state. A common observation of this effect is in the J and H aggregates of chromophores; a simple visualization involves the transition dipoles of the interacting molecules in a head-to-tail fashion in the former, and side-by-side fashion in the latter. The coupling leads to red- and blue-shifted optical absorptions respectively in these aggregates.

6.1.4 Raman Scattering

Interaction of photons with materials can also lead to scattering. Scattering is essentially the deviation of light from its path, because of the inhomogeneity in the path due to molecules or materials. A scattering that occurs without exchange of energy between the radiation and the material is called elastic; the incident and scattered photons have the same energy, or the wavelength of light does not change in this process.

- Scattering of light by objects much smaller than the wavelength of the light (atoms and molecules) is known as **Rayleigh scattering**. The intensity of scattering is strongly dependent on the wavelength, $I \propto \lambda^{-4}$, becoming dominant at short wavelengths; this effect contributes to the blue color of the sky.
- **Mie scattering** arises when objects have sizes similar to or larger than the wavelength of the light. The scattering does not depend strongly on the wavelength; the white color of clouds is attributed to the scattering by water droplets.

Besides Rayleigh scattering, light scattered from molecules and crystals may contain photons with different energy (or wavelength). A well-known inelastic scattering of light (energy exchanged with the medium) is **Raman scattering**. Raman spectroscopy is an important tool to study the structure of crystals.

- Raman scattering with incident and scattered photons of different energies involves the creation or annihilation of phonons. The selection rules are: $\omega = \omega' \pm \Omega$ and $\mathbf{k} = \mathbf{k}' \pm \mathbf{K}$, where ω and $\mathbf{k}$ are the frequency and wave vector of the incident photon, ω' and $\mathbf{k}'$ refer to the scattered photon, and Ω and $\mathbf{K}$ refer to the phonons; $\pm$ represent the creation/annihilation of the phonons.
- These processes are represented schematically in Figure 6.3. In a Raman spectrum, the peaks due to the scattered photons with lower frequency (resulting from creation of phonons) are known as the **Stokes lines**, and those with higher frequency (resulting in annihilation of phonons) are known as **anti-Stokes lines**. Anti-Stokes lines vanish at low temperature as phonons are fewer.
- When the phonon involved is an acoustic phonon, it is known as **Brillouin scattering**.

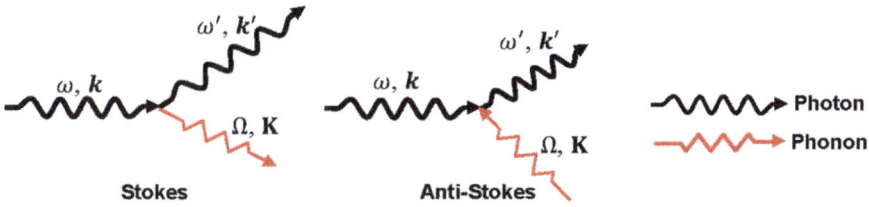

Figure 6.3 Raman scattering of a photon with emission (Stokes) or absorption (anti-Stokes) of a phonon.

6.1.5 Electron Spectroscopy

Another level of interaction of photons with materials involves the ejection of electrons from the material. When photons with the appropriate energy ($h\nu$) are incident on the material, electrons are ejected.

- The binding energy (BE) of the electrons in the material can be estimated from the energy conservation equation: $BE = h\nu - KE$, where KE is the kinetic energy of the ejected electrons.
- The ejected electrons are known as photoelectrons and the spectroscopy developed based on this principle is photoelectron spectroscopy.
- Information about the electron binding energies, gives precise information about the nature of the atoms present, their oxidation states, and the environment. They provide insight into the band structure of the solid.

Depending on the energy of the photons employed, the technique is known as ultra-violet photoelectron spectroscopy (UPS) or X-ray photoelectron spectroscopy (XPS).

- UPS involves UV photons with energy in the range 10–45 eV ($\sim$124–27 nm); it is useful to probe valence electrons in the atoms.
- XPS involves X-ray photons with energy in the range 200–2000 eV, and are used to probe the core electrons.
- The photoelectron spectroscopy primarily probes atoms close to the surface of a material, down to a depth of about $\sim$50 Å.

6.2 Photoconduction

When light is absorbed by a material, the intensity, I decreases exponentially with the path length, x within the material, as $I = I_0 e^{-kx}$ where I_0 is the incident light intensity and k is the absorption coefficient. Free carriers are generated in the valence band, when light with frequency, ν is absorbed by intrinsic semiconductors (band gap, E_g) or extrinsic semiconductors (impurity level gap, E_{imp} (E_D or E_A)), if the following condition is satisfied:

- $h\nu \geq E_g$ for an intrinsic semiconductor $\left[\lambda \leq \dfrac{hc}{E_g} \right]$
- $h\nu \geq E_{imp}$ for an extrinsic semiconductor $\left[\lambda \leq \dfrac{hc}{E_{imp}} \right]$

The conductivity in the absence of light is due to thermal excitation of charge carriers; this is called the '**dark conductivity**'. Free charge carriers produced upon excitation by light, leads to an increase in the conductivity; this additional conductivity is known as '**photoconductivity (σ_{ph})**'. In materials with large exciton binding energies, the thresholds for optical absorption and photoconductivity are different.

- Charge carrier excitation across the band gap and hence photoconduction, in intrinsic semiconductors ($\sim$1–3 eV) is possible by visible or near infra-red light.
- In extrinsic semiconductors with small energy gaps ($\sim$0.1 eV), photoconduction is possible by infra-red light irradiation. Since the impurity atoms are fully ionized above the exhaustion temperature, T_S (Section 4.3.5), and hence no photoexcitable electrons are present in the impurity levels, the extrinsic semiconductors need to be cooled below T_S to observe photoconduction by infra-red irradiation.
- The typical conductivity response of photoconductors is shown schematically in Figure 6.4(a). When the irradiation is switched on, the photoconductivity rises to reach an equilibrium value;

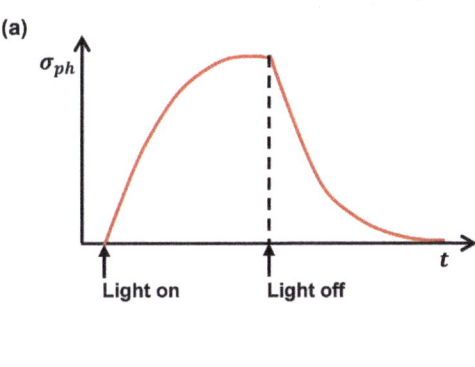

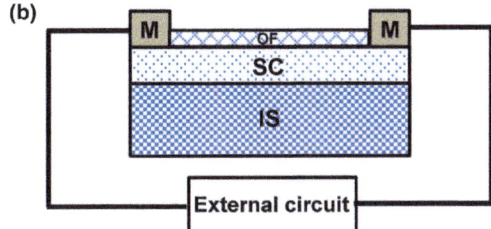

Figure 6.4 (a) Schematic diagram of the photoconductivity response as a function of time, with light on and off leading to photoconductivity rise and fall. (b) Schematic diagram of a photoresistor described in the text.

upon switching the irradiation off, it decreases exponentially, $\sigma_{ph} = \sigma_{ph,0} e^{-t/\tau}$ where τ is the excess charge carrier lifetime.

6.2.1 Photoresistor

Photoconduction is widely used in photoresistors, employed in light-sensitive detector circuits, and light/dark-activated switching circuits.

- They are made of high resistance (dark resistance) semi-conductors that show significant increase in conductivity upon light irradiation; for example, CdS commonly used in photresistors has a $\dfrac{\sigma_{ph}}{\sigma_{dark}} \sim 10^5 - 10^6$.
- Typically, the semiconductor in the form of a thin film (SC) is deposited on an insulating substrate (IS), connected to an external circuit using metal electrodes (M) and protected by a transparent organic film (OF) (Figure 6.4(b)).

6.2.2 Xerography

The vital component of xerography, or the process of xeroxing (originally called electrophotography), is a photoconductor. The primary component of the xerox machine is a metal cylinder called the **drum**, coated with a photoconductor; in the early machines, amorphous selenium was the photoconductor, however, ceramic and organic photoconductors are used in recent times. With organic photoconductors, flexible oval or triangular belts can be used in place of the drum, allowing smaller device size. The main steps involved in xeroxing (Figure 6.5) are:

- **Charging:** The drum surface is charged, usually negatively, by discharge from a wire.
- **Exposure:** The document to be copied is illuminated by a flash lamp and the image projected on to the drum surface (which may be rotated, synchronized with the scanning of the document). Where the light falls, the surface charge is carried away (thanks to the photoconductivity) and where the image falls, the surface charge remains.
- **Development:** Spraying the drum surface with the toner (usually containing carbon powder, and statically charged with a charge opposite to that on the drum surface), causes the toner to collect together and stick to where the surface charge remains, thus developing the image of the original document.

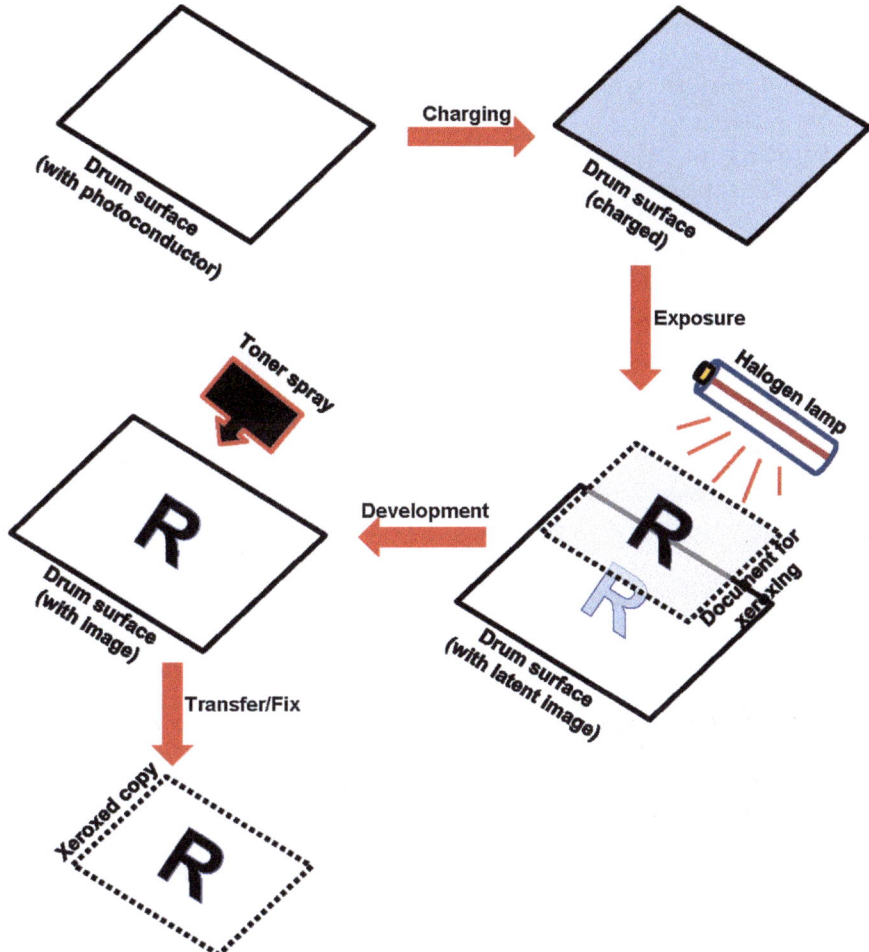

Figure 6.5 Schematic illustration of the basic steps involved in the process of xerography.

- **Transfer and fixing:** Using pressure and electrostatic force, the toner image on the drum surface is transferred on to a paper. Using heat and pressure, the image is fixed on the paper. The drum is cleaned and used in another cycle.

6.2.3 Photography

Even though conventional photography has largely disappeared now, it illustrates the versatile utility of photoconductivity and ionic conductivity of silver halides. The basic steps involved in conventional photography are the following.

- Exposure of the film containing silver halide to light from the subject being photographed.
- Development and fixing of the latent image to form the 'negative'.
- Printing of the 'positive', the real image of the object on a photographic paper.

In the first stage of **exposure**, the silver halide is converted to silver at the points where light falls on the film. The mechanism involved in this process is possibly quite complex; the model proposed originally by Gurney and Mott[2] involves photoconductivity of silver halide under visible light irradiation. The mechanism can be summarized as follows, taking the example of AgBr as the photoactive material:

- Irradiation of AgBr grains produces an electron–hole pair.
- Thanks to the photoconductivity of AgBr, the electron (in the conduction band due to photoexcitation) migrates to the grain surface, where it is trapped by impurities such as Ag_2S.
- The trapped electron attracts Ag^+ ions from the Frenkel defects in the AgBr (see Section 2.2.4) to form an Ag atom.
- The Ag atom combines with another trapped electron to form Ag^- which in turn captures a second Ag^+ to form Ag_2.
- This process continues to form a colloidal particle of silver at the points where light was incident on the film, to form the latent image.
- The Br^- ions in the lattice possibly capture the photo-generated holes to form Br atoms and subsequently Br_2, which escapes as a gas: $Br^- + h^+ \rightarrow Br \rightarrow \frac{1}{2}Br_2\uparrow$.

The latent image is **developed** using chemicals like hydroquinone and washed in hypo solution (ammonium thiosulphate) to remove the unreacted silver halide in the un-irradiated regions; the irradiated regions with silver remain, to form the 'negative'. The **printing** involves irradiation of the negative, which provides a positive image on the photographic paper, as light passes through the regions where there is no silver.

6.3 Luminescence

Atoms, molecules, and materials, excited electronically using appropriate energy sources, can undergo a variety of photophysical and photochemical processes from the excited state. One of the important photophysical processes is luminescence, which involves the excited state decay through emission of energy in the form of light.

If the electronic excitation is carried out by light (photoirradiation), the subsequent emission is known as photoluminescence. When the excitation is effected by the application of an electric field, it is known as electroluminescence.

6.3.1 Photoluminescence

Fluorescence: When atoms or molecules are irradiated by UV or visible light, they can be promoted to their electronically excited singlet state, S_1 (or to S_n which due to fast internal conversion decays to the S_1 state); it can subsequently decay back to the ground state (S_0), emitting light. This process occurs typically over 10^{-10}–10^{-7} s (nanosecond time scale), and is called fluorescence. As shown in Figure 6.6, the emitted light usually has a longer wavelength (smaller energy photons) compared to that of the light absorbed. This arises due to the relaxation of the molecule down the vibrational levels in the excited state. The energy shift of the emission with respect to the excitation is called the **Stokes shift**.

Phosphorescence: The molecule in the excited singlet state (S_1) can decay to a triplet state (T_1) *via* intersystem crossing, which subsequently can radiatively decay back to the ground state (Figure 6.6), emitting photons of much lower energy than those absorbed.

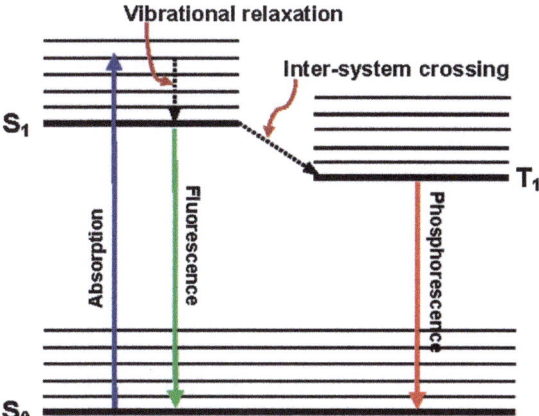

Figure 6.6 A simplified Jablonski diagram showing two singlet and one triplet electronic states (thick horizontal lines) along with their vibrational levels (thin horizontal lines), absorption, fluorescence and phosphorescence transitions, as well as vibrational relaxation and inter-system crossing processes.

Intersystem crossing between the singlet and triplet states is spin-forbidden, but is facilitated when spin–orbit coupling is strong. As the $T_1 \rightarrow S_0$ transition is slow, the light emission (phosphorescence) occurs typically over 10^{-5}–10^3 s.

Many molecules and materials show strong luminescence which is of great utility in applications such as sensors, display systems, imaging microscopy, *etc.* Selected examples of such materials are the following:

- Organic molecules such as pyrene and dyes like cyanines, fluorescein, and rhodamine as well as conjugated polymers like poly(*p*-phenylenevinylene) are well-known examples of fluorescent systems.
- Several metal complexes (*e.g.*, ruthenium with bipyridine and phenanthroline based ligands), as well as complexes of inner transition metal ions like Tb^{3+} also show strong fluorescence. Ir(III) complexes are well-known examples of phosphorescent molecules.
- Many semiconductors show strong photoluminescence. The electronic excitation from the valence band creates excitons. Radiative decay of the excitons leads to light emission. In conventional semiconductors, the exciton binding energy is very small, and the excitation and de-excitation occur from band states. If the semiconductor is a direct band gap one, both excitation and radiative decay are allowed and phonons are not involved. Hence light emission from direct band gap semiconductors is more efficient than from indirect band gap semiconductors (Section 4.3.2). Popular examples of semiconductors that exhibit photoluminescence include ZnO, ZnS, CdS, CdSe, GaN, $CuInS_2$, *etc.*

6.3.2 Special Mechanisms of Fluorescence Emission

As noted above, fluorescence is associated with a Stokes shift; the emitted photons have lower energy than the exciting photons. Figure 6.6 shows that this results from the vibrational relaxation in the excited state. It is also possible that an excited state reached by photon absorption, decays to a lower lying excited state from which radiative emission occurs. This process is called **down-conversion**. In some cases, the absorption at one site is followed by energy transfer to another site from which down-converted photon emission occurs (Figure 6.7(a)).

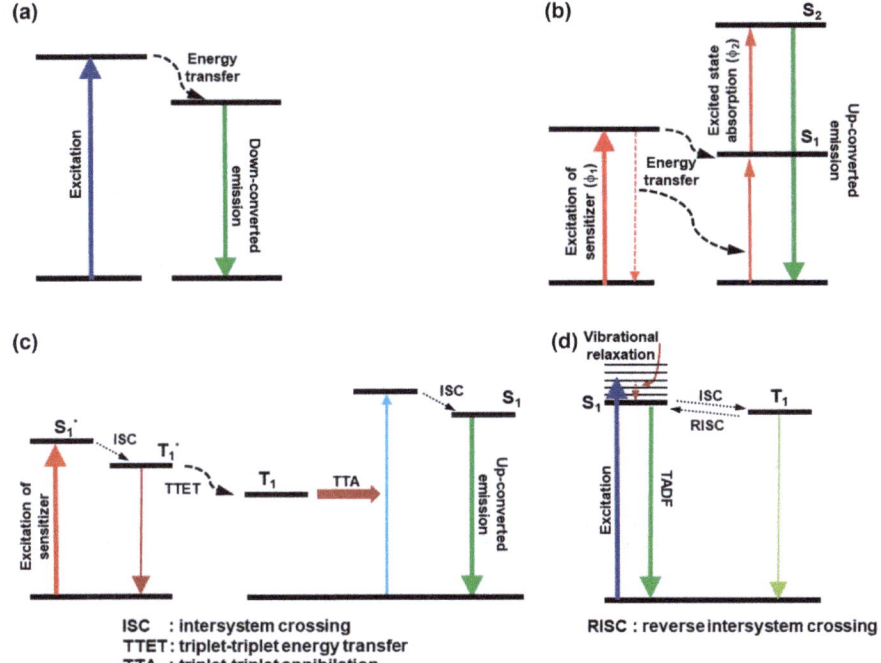

Figure 6.7 Schematic of typical energy level diagrams and photophysical processes occurring in (a) down-conversion luminescence, (b, c) up-conversion luminescence (the latter based on triplet–triplet annihilation (TTA)), and (d) thermally activated delayed fluorescence (TADF). The lowest energy states are the singlet (S_0) in all cases; S_1, S_2 are the excited singlet states and T_1 the triplet state. Vertical arrows: colors are approximate representations of the relative energies, and thickness, approximate representation of the relative intensities.

It is possible to achieve anti-Stokes shifted luminescence under special conditions, through an **up-conversion** process. One mechanism of up-conversion involves excitation of a luminescent center from the ground electronic state to excited state **1** (with a relatively long lifetime), followed by its excitation to state **2**. Photon emission with energy equal to the sum of the two excitations, then occurs from excited state **2**.

- The excitations can be by direct photon absorption, or energy transfer from a sensitizer as shown in Figure 6.7(b).
- This process is common in lanthanide based up-conversion systems.

Another mechanism of up-conversion uses a sensitizer excited to its triplet state through intersystem crossing, and energy transfer to the emitting center occurs, followed by the process of **triplet–triplet annihilation** (a typical scheme is shown in Figure 6.7(c)).

- The energy transfer from the triplet state of the sensitizer to the triplet state of the emitter leads to the formation of a large population of the long-lived excited triplet state of the emitter.
- Addition of the energy of two triplets enables the excitation of a higher energy singlet state; this process is known as **triplet–triplet annihilation**.
- Emission of the up-converted photons occurs from this singlet state.
- Several polycyclic aromatics and heterocyclic compounds are well-known emitters, and porphyrin and phthalocyanine based complexes, are efficient sensitizers.

Thermally activated delayed fluorescence (TADF) is a process in which triplet states (which do not emit fluorescence as discussed earlier), use thermal energy from the surroundings and undergo a reverse intersystem crossing (RISC) to the close-lying singlet excited state. Fluorescence occurs from the singlet excited state (Figure 6.7(d)).

- The longer fluorescence time-scale results from the multiple steps involved.
- Many such systems show a prompt fluorescence in addition to delayed fluorescence emission.
- A small energy gap between the excited singlet and triplet states (Δ_{ST}) facilitates TADF emission.
- A popular design of TADF emitters uses donor–acceptor systems with a strong dihedral twist between them, leading to near degeneracy of the singlet and triplet states. Carbazoles are common donor moieties and triazines, benzophenones, *etc.*, are common acceptor groups.

6.3.3 Electroluminescence

An electronically excited state of molecules or excitons in semiconductors can also be created by the application of a suitable electric field.

- In a molecule (semiconductor) sandwiched between an anode and a cathode (Figure 6.8(a)), the process involves the injection of an electron into the LUMO (conduction band) from the cathode and injection of a hole into the HOMO (valence band) from the

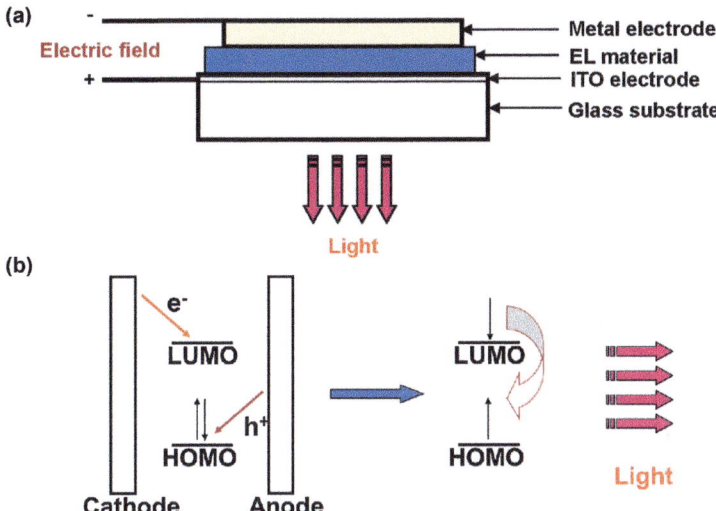

Figure 6.8 Schematic diagrams showing (a) light emission from an electroluminescent (EL) material sandwiched between a cathode (metal electrode) and an anode (transparent indium tin oxide (ITO) on glass) subjected to an electric field, and (b) the process of creation of excited state by the applied electric field, which decays by light emission.

anode. The process is shown schematically for a molecule, in Figure 6.8(b).

- This results in the formation of the electronically excited state of the molecule (exciton in solids), which decays radiatively, producing light emission.
- This is the principle of light emitting diodes (LEDs); popular materials used in LEDs include InGaN (blue, green, and UV), AlGaInP (yellow, orange, and red), and AlGaAs (red and infra-red). AlQ$_3$ ((*tris*-8-hydroxyquinoline)aluminum), conjugated polymers based on poly(*p*-phenylenevinylene), *etc.*, are important examples of electroluminescent molecular materials. LEDs based on organic emitters are called OLEDs.

6.3.4 Other Forms of Luminescence

Besides photoluminescence (triggered by light absorption) and electroluminescence (induced by an electric field), luminescence can occur by absorption of energy by materials, through a variety of other routes. Some of the more prominent ones are noted below.

Thermoluminescence occurs when materials which have been electronically excited by electromagnetic or ionizing radiations previously, are heated. The excited states are trapped in defects or imperfections in the lattice and the heating induces their decay emitting light.

- Proportionality of the thermoluminescence response to the radiation dose received by a material, is used in thermoluminescence dating of the age and history of the material.
- Fluorite is a typical mineral that is known to exhibit thermoluminescence.
- Thermoluminescence should be distinguished from **incandescence**. The latter refers to the emission of electromagnetic radiation (including in the visible range) when a substance is heated. The radiation wavelength depends only on the temperature to which it is heated (recall black body radiation). Production of light in the common tungsten filament bulbs is based on incandescence.

Mechanoluminescence results from the application of mechanical force on a material. The historically famous example is the sparkling light produced when hard sugar is scraped with a knife. The luminescence is known to result from either deformations of the substance or just rubbing; the latter is known as **triboluminescence**.

- The former occurs due to physical processes induced during the deformation, for example, fracture of the material and surface charges due to piezoelectric effects. The ensuing discharge can excite the molecules on the surface or in the surrounding gas leading to light emission.
- Triboluminescence is due to contact phenomena such as triboelectricity and tribochemical reactions when dissimilar materials are contacted or separated.
- Examples of mechanoluminescent materials include minerals like quartz, rare earth metal ion complexes and organic compounds like some anthracene derivatives.

Chemiluminescence is the emission of light from the excited state of a molecule formed (not by photoexcitation as in fluorescence or phosphorescence) as the product of a chemical reaction, $A + B \rightarrow C^* \rightarrow C + h\nu$ (light).

- The classic example is luminol treated with a base and an oxidizing agent like hydrogen peroxide in the presence of a catalyst, typically ferricyanide or metal ions.

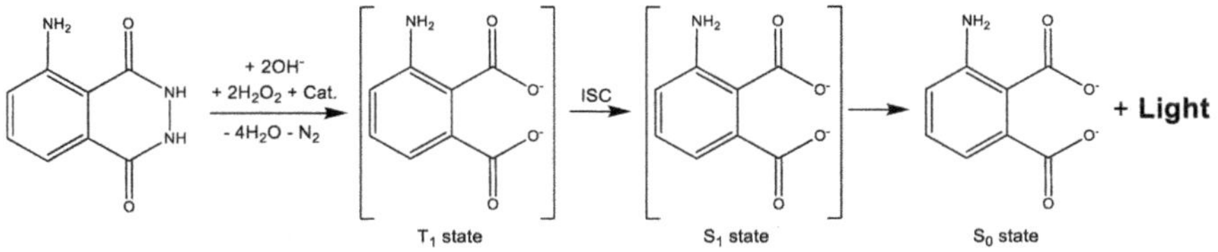

Figure 6.9 Mechanism involved in the chemiluminescence of luminol.

- A general mechanism involves the following steps (Figure 6.9). The product is formed in the triplet excited state (T_1), which undergoes intersystem crossing to the singlet excited state (S_1), which emits light and decays to the ground electronic state (S_0).
- This reaction forms the basis for the luminol test for blood at crime scenes; iron in the hemoglobin acts as the catalyst producing the chemiluminescence for detection.

Bioluminescence is essentially chemiluminescence occurring in a living organism, mediated by suitable enzymes.

- It occurs in many marine vertebrates and invertebrates, and in microorganisms and terrestrial species like fireflies and glowworms.
- A common mechanism of bioluminescence, for example that occurring in fireflies, involves the light emitting pigment luciferin and the enzyme luciferase. Formation of oxyluciferin in presence of oxygen is accompanied by light emission.

$$\text{Luciferin} + O_2 \xrightarrow{\text{Luciferace+Cofactors}} \text{Oxyluciferin} + \textbf{Light}$$

6.4 Lasers

A **laser** (**l**ight **a**mplification by **s**timulated **e**mission of **r**adiation) is based on the phenomenon of stimulated light emission; the difference with respect to spontaneous emission which leads to luminescence, is illustrated in Figure 6.10.

- Rate of spontaneous emission from an excited state to the ground state $= -\dfrac{\partial N_2}{\partial t} = A_{21}N_2$ where N_2 is the population of the excited state and A_{21} is the Einstein A coefficient.
- Rate of stimulated emission $= -\dfrac{\partial N_2}{\partial t} = B_{21}\rho(v)N_2$ where N_2 is the population of the excited state, B_{21} is the Einstein B coefficient and

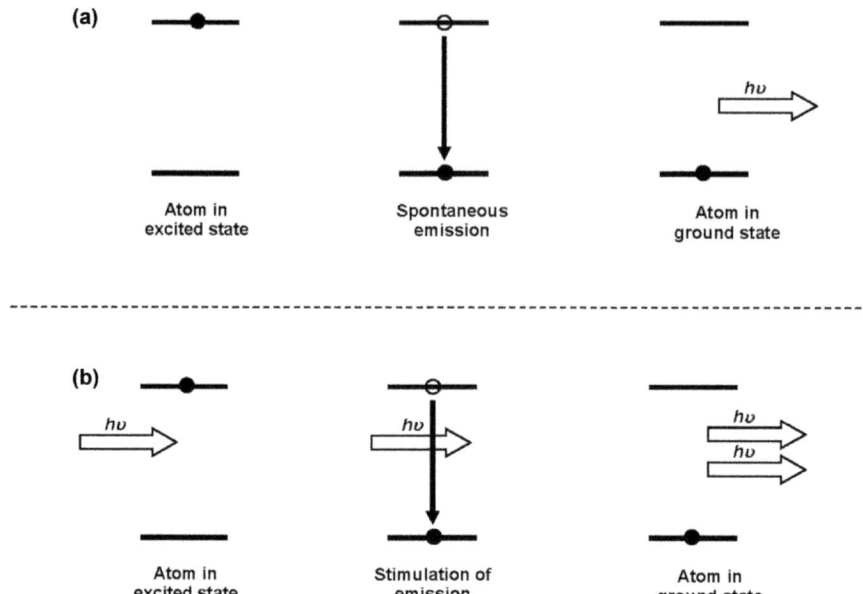

Figure 6.10 Schematic representation of (a) spontaneous emission (fluorescence, phosphorescence, *etc.*), and (b) stimulated emission (lasers) processes.

$\rho(v)$ is the radiation density of the incident field at frequency v. The emitted photons follow the phase of the stimulating photons.
- Lasing involving stimulated emission has the following characteristic features. It is monochromatic (single wavelength), coherent (all photons in phase) and directional (scattered very little).

6.4.1 Population Inversion

Under thermal equilibration (at temperature, T), the population in the excited state (N_2) is lower than that in the ground state (N_1); ratio of the populations is given by the Boltzmann equation, $\dfrac{N_2}{N_1} = e^{-\Delta E/k_B T}$ where ΔE is the energy difference between the two states. Stimulated emission can be enhanced and made dominant, if the population of the excited state is higher than that in the ground state, *i.e.*, if there is a '**population inversion**'.

- One way of achieving an inverted population makes use of a careful balance of transition probabilities between different states of the gain medium.

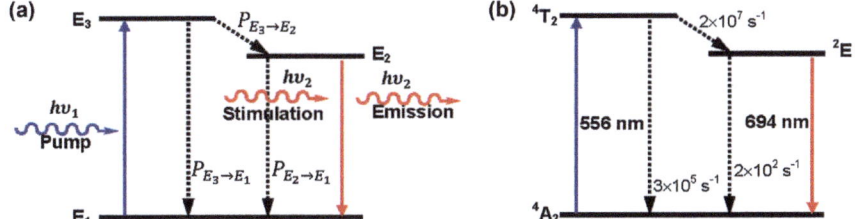

Figure 6.11 Schematic diagrams of: (a) a 3-level system to generate population inversion between states E_1 and E_2 (the pump, stimulation and laser emission are shown), and (b) the electronic states and transition rates of Cr^{3+} doped in alumina (ruby laser).

- Consider a 3-level system (Figure 6.11(a)). If the atom (or molecule, material, *etc.*) is excited from state E_1 to E_3 (known as pumping), and the transition probabilities (P) are such that, $P_{E_3 \to E_2} \gg P_{E_3 \to E_1}$, then most of the excited atoms will funnel down to E_2. If $P_{E_3 \to E_2} \gg P_{E_2 \to E_1}$ also, then the excited atoms pile up in E_2. With sufficient power used for the $E_1 \to E_3$ transition, population inversion between E_1 and E_2 is achieved.

- A typical example is the case of a ruby laser, which uses Cr^{3+} ions doped in alumina (Al_2O_3), the Cr^{3+}/Al^{3+} ratio being ~1/2000. Alumina is colorless; the ruby red color arises due to the visible absorption involving Cr^{3+} levels lying within the band gap of alumina. The absorption occurs at 556 nm due to the $^4A_2 \to {}^4T_2$ transition (Figure 6.11(b)). The rapid $^4T_2 \to {}^2E$ radiation-less transition (rate $= 2 \times 10^7$ s^{-1}) compared to the $^4T_2 \to {}^4A_2$ transition (3×10^5 s^{-1}), coupled with the slow $^2E \to {}^4A_2$ emission (2×10^2 s^{-1}) leads to enhanced population of ^2E and hence population inversion.

- 4-level systems are even better suited to achieve population inversion.

- The ruby laser discussed above and well-known systems like the Nd:YAG laser, use solid state gain media. Lasers based on gas and liquid media are also popular. Examples of the former are helium–neon, argon ion, and carbon dioxide lasers. Dye lasers have a liquid gain medium; dyes like rhodamine 6G are commonly used, and the solutions in organic solvents are pumped through a flow cell.

6.4.2 Optical Resonator (Cavity)

After the inverted population is achieved, intense laser emission is generated using a resonating cavity. The cavity design and steps involved in the process are as follows.

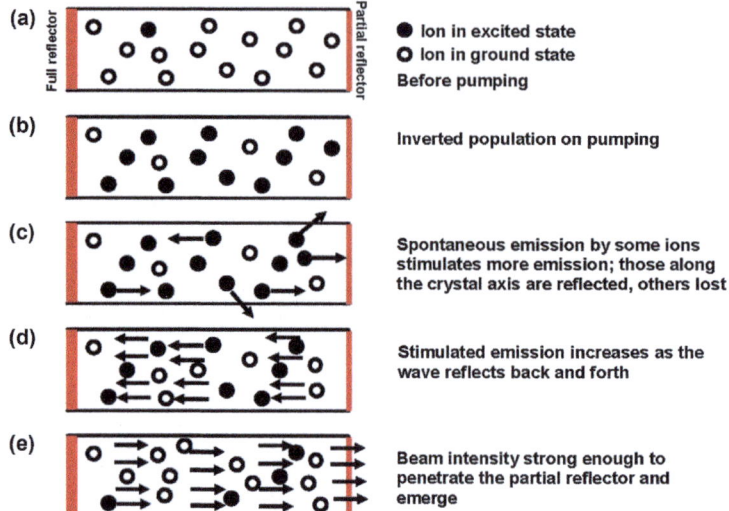

Figure 6.12 Stages in the lasing process: (a, b) pumping and population inversion, (c, d) intensity build-up, and (e) beam generation in an optical resonator (cavity); see text for details of each stage.

- An optical resonator or cavity has two parallel mirrors placed on either side of the lasing medium, one highly reflective and the other partially reflective. This can be achieved by polishing parallel faces of a crystal and coating them with silver, such that one is fully opaque, while the other is partially transparent; most of the atoms or ions in the lasing medium are initially in the ground state (Figure 6.12(a)).
- After pumping, an inverted population is achieved (Figure 6.12(b)). Photons emitted spontaneously by a few excited atoms/ions, along the crystal axis stimulate more to emit photons along the same direction (Figure 6.12(c)).
- Stimulated emission increases as the building wave reflects back and forth between the mirrors (Figure 6.12(d)); each excited atom/ion emits its energy and the wave fronts of the stimulated photons are in phase with the propagating beam. When the wave intensity become strong enough to penetrate the partially reflective mirror, the laser beam emerges (Figure 6.12(e)).

6.4.3 Diode Laser

Known also as an **injection laser** or **laser diode**, this is commonly used in optical fiber applications, laser printers, and compact disk systems.

They are based on semiconductor p–n junctions (Section 4.3.6); direct band gap semiconductors are used. Cavity design facilitates stimulated emission to occur, as opposed to the spontaneous emission in LEDs (Section 4.3.3).

- Figure 6.13(a) shows the band bending at the p–n junction that leads to equalization of the Fermi levels, and build-up of the energy barrier qV_0 (q = charge) that stops the diffusion of electrons and holes across the junction.
- Application of a forward bias (applied potential, V so that $qV > E_g$, the band gap) across the p–n junction shifts the bands so that a population inversion is created (Figure 6.13(b)); stimulated emission follows, as indicated.
- The diodes are polished so that one side is a fully reflecting mirror and the other side is partially reflecting; this leads to intensity build-up and lasing as discussed in Section 6.4.2.
- Different semiconductors can be chosen to obtain different emission wavelengths; for example, AlGaAs (720–850 nm), AlGaInP (650 nm), InGaN (405 nm), *etc.*

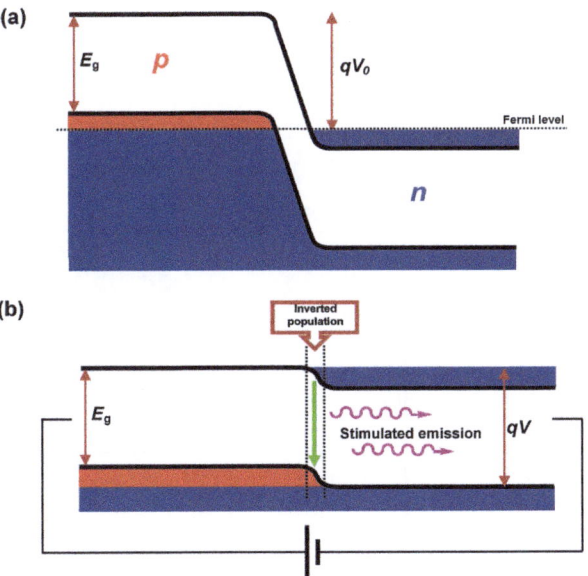

Figure 6.13   Schematic diagrams of: (a) the p–n junction with energy barrier, qV_0, and (b) the band shift under positive bias, qV that leads to the diode laser action. Red: holes; blue: electrons.

6.5 Photovoltaic and Photoelectrochemical Effects

In Section 6.3.3, the phenomenon of electroluminescence was discussed; application of an electric field across an appropriate material generates excitons, decay of which leads to light emission. The reverse process produces the **photovoltaic effect**: irradiation of a material with light of a suitable wavelength creates excitons, which in turn are split into electrons and holes and collected at the appropriate electrodes, generating a voltage. It thus converts light energy to electrical energy. If there is also an electrochemical process coupled with this, it is called **photoelectrochemical effect**.

6.5.1 Photovoltaic Cells

A p–n junction irradiated with light of suitable frequency creates enhanced population of holes and electrons on the p- and n-regions respectively (Figure 6.14). This is effectively similar to the application of a forward bias (Figure 4.22), lowering the potential gradient across the junction; the resulting photovoltage can drive a current in an external circuit.

- Efficiency of conversion of light energy to electrical energy is expressed as, $\eta = \dfrac{\text{output power } (I \times V)}{\text{input power (light)}}$

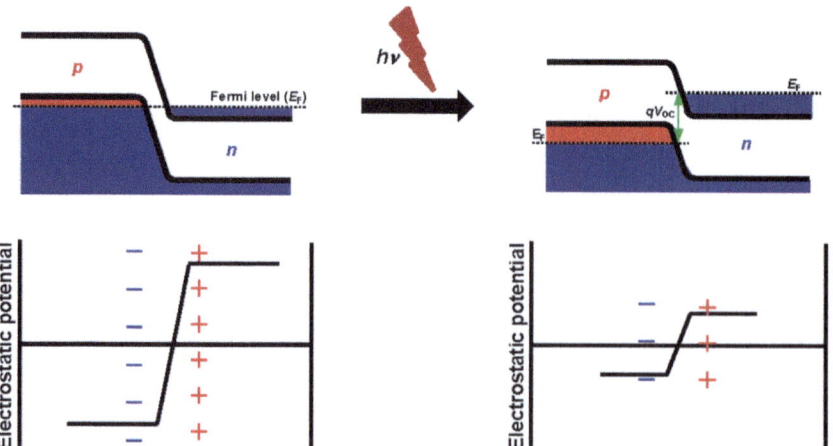

Figure 6.14 Schematic diagram of the band structure and filling at the p–n junction and its change upon illumination, leading to the generation of the photo voltage (open-circuit potential, V_{OC}); the potential gradient changes are also shown.

- Examples of early photovoltaic materials: Si ($E_g = 1.1$ eV, $\eta \sim 20\%$), GaAs ($E_g = 1.35$ eV, $\eta \sim 12\%$), CdSe ($E_g = 1.74$ eV, $\eta \sim 5\text{-}7\%$).
  Several generations of photovoltaic cells have emerged over the years.
- Si based solar cells are still dominant at the commercial level. They come in the forms of monocrystalline silicon, polycrystalline silicon, and amorphous (thin film) silicon; the efficiency decreases in the same order. Thin film silicon can be prepared on different substrates including glass.
- Solar cells based on CdTe and CuInGaSe are also popular.
- Another class of photovoltaics are based on organic materials, that utilize the idea of donor–acceptor systems and heterojunctions. A typical example is the composite made using various derivatives of poly(*p*-phenylene vinylene) (PPV) serving the role of donors, with fullerenes as the acceptor.
- Dye-sensitized solar cells (DSSC), the Grätzel cell being the most popular design, use a semiconductor anode of TiO_2 and photosensitizers such as ruthenium dyes adsorbed on TiO_2. The photoexcited electrons are channeled into the semiconductor anode. The holes are scavenged using an iodide/triiodide (I^-/I_3^-) redox system; the triiodide gets reduced at the counter electrode.
- Several new generations of light absorbing semiconductors have been developed as suitable candidates for increasingly more efficient photovoltaics. Perovskite materials have emerged as one of the most successful.

Photodiodes are essentially photovoltaic devices; often they are used in reverse bias mode. As the photocurrent generated is proportional to the intensity of light, they are useful as photo-detectors.

6.5.2 Photoelectrochemical Cells

A photoelectrochemical cell is based on the absorption of light coupled with electrochemical events that then lead to the production of a voltage. DSSC mentioned in the previous section is an example of this kind. The working principle is illustrated in Figure 6.15. Consider an n-type semiconductor electrode placed in contact with a redox system in an electrolyte.

- The valence and conduction bands of the semiconductor bend due to the flow and accumulation/depletion of charge inside the electrode and of ions (*e.g.*, H^+ or OH^-) on the solution side.

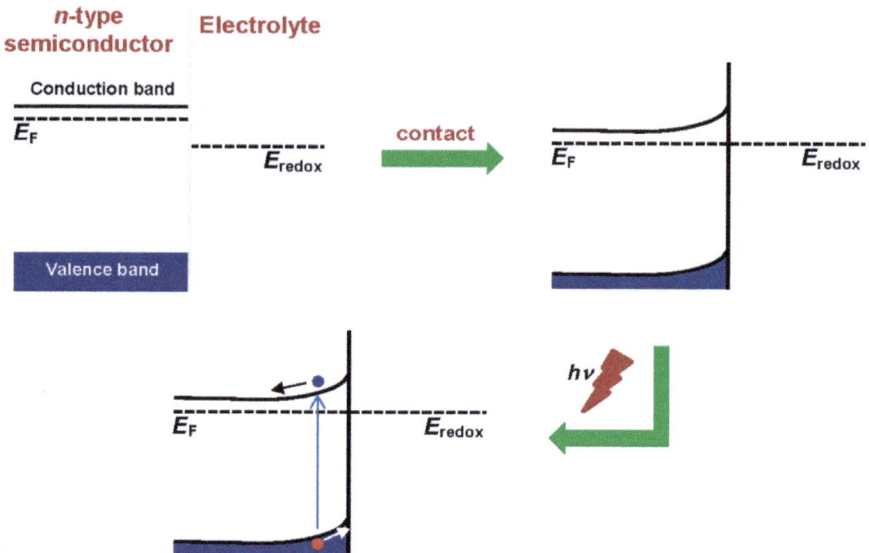

Figure 6.15 Schematic band diagram of an n-type semiconductor, bending of the bands when brought in contact with a redox system, excitation of electron–hole (blue–red) pair and its separation into the electrode and the electrolyte interface.

This can be viewed as resulting from the equilibration of the Fermi level of the semiconductor with the redox level of the electrolyte.

- Photoirradiation leads to the excitation of electrons from the valence to the conduction band. The strong field at the contact separates the electron–hole pair, the electron is swept into the electrode and the hole into the electrolyte. Note that electrons move downwards and holes upwards on the energy scale.

The charge separation produces a counter field that shifts the band and creates the photovoltage, which can be effective in carrying out electrolysis of the redox system. A counter electrode is needed for the reduction to occur when oxidation occurs at the photo-anode.

A typical example is a photoelectrochemical cell with an *n*-type photo-anode and a p-type photo-cathode. Figure 6.16 shows the electrolysis of water using such a system. If the band gap is 1.5–2.0 eV and the Fermi levels of the electrodes are appropriate, the following reactions involved in the water splitting occur.

- Anode (n-type): $4h^+ + 2H_2O(l) \rightarrow 4H^+ + O_2(g)\uparrow$ ($E^\circ = 1.23$ V *versus* RHE)

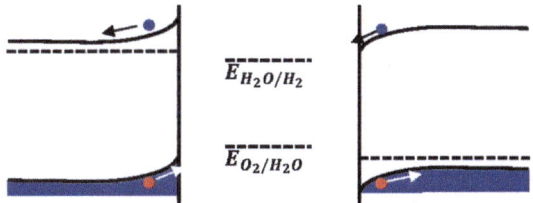

Figure 6.16 Illustration of n-type and p-type semiconductors acting as the anode and cathode respectively in a photoelectrochemical cell; hole (red) from the valence band of the anode oxidizes water to oxygen, and electron (blue) from the conduction band of the cathode reduces water to hydrogen.

- Cathode (p-type): $4e^- + 4H^+ \rightarrow 2H_2(g)\uparrow$ ($E^\circ = 0.0$ V *versus* RHE)
- The total reaction can be written as: $2H_2O(l) \xrightarrow{h\nu} 2H_2(g)\uparrow + O_2(g)\uparrow$

6.6 Basic Principles of Nonlinear Optics

Some of the common optical processes involving materials are reflection, refraction, and transmission/absorption. During these light–matter interactions, generally, neither the properties of the material medium (such as refractive index, absorption coefficient), nor the properties of light (such as frequency) change. Such processes are called linear. With the advent of lasers, and the high intensities that such light possesses, it was found that the light–matter interaction can significantly alter the properties of the material and the light, during the interaction. Such processes are called nonlinear; the media are nonlinear optical (NLO) materials. Typical examples are the following.

- The refractive index, and therefore the speed of light in the NLO medium can vary with the intensity of the light.
- The frequency of light can change as it passes through the NLO material; for example it can change color from red to blue.
- Photons can interact within the NLO material; the interaction mediated by the material, allows the control of light by light. This is one of the basic approaches to the emerging field of photonics.

6.6.1 Linear and Nonlinear Polarization Response

It was mentioned in Section 6.2 that the polarization, $P(\omega)$ is proportional to the electric field, $E(\omega)$ of the electromagnetic wave with frequency, ω. However, with intense light, the polarization will have a

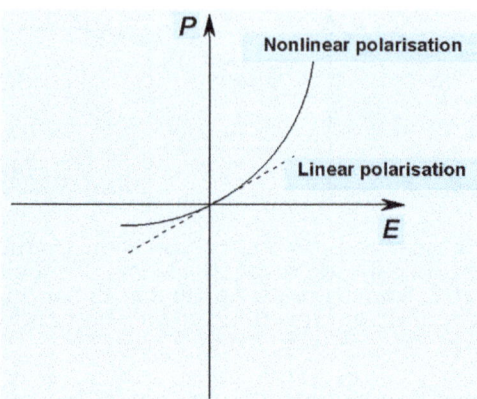

Figure 6.17 Schematic plot of the nonlinear variation of polarization (*P*) with electric field (*E*); the linear response at small values of *E* is indicated by the broken line. The frequency and vector component symbols are omitted for simplicity.

nonlinear dependence on the field; the case with a medium lacking a center of inversion is shown in Figure 6.17. The nonlinear effects become visible, when $E(\omega)$ is comparable to or stronger than the interatomic electric fields.

This can be understood as follows. When the electric field is small, the equilibrium charge displacement is proportional to the field, since the restoring elastic force is proportional to the displacement (Hooke's law). When the field is very high, the restoring force is a nonlinear function of the displacement, leading to nonlinear polarization.

Noting that polarization and electric field are vector quantities, the ith component of the induced nonlinear polarization resulting from incident intense light of frequency ω, can be written as a function of the electric field components, *j*, *k*, *etc.*, as follows:

$$P_i(\omega) = \chi_{ij}^{(1)} \, E_j(\omega) + \chi_{ijk}^{(2)} \, E_j(\omega)E_k(\omega) + \chi_{ijkl}^{(3)} \, E_j(\omega)E_k(\omega)E_l(\omega) + \dots$$

- In the above expression, $\chi_{ij}^{(1)}$ is the linear electric susceptibility tensor coefficient; it represents the linear response of the polarization with the electric field. $\chi_{ijk}^{(2)}$, $\chi_{ijkl}^{(3)}$, *etc.*, are the nonlinear susceptibility tensor coefficients representing second order, third order, *etc.*, effects.
- Note that tensors relate two or more vectors. For example, the quadratic susceptibility tensor, $\chi^{(2)}$ relates the *P* vector (row vector: $i = x, y, z$) with the combinations of two *E* vectors (column vector: $jk = xx, xy, xz, yx, yy, yz, zx, zy, zz$); it will be a $3 \times 3 \times 3$ object

having 27 components ijk $(i, j, k = x, y, z)$, which for convenience is generally represented as a 3×9 matrix.

- The magnitude of susceptibility coefficients decrease with increasing order. Hence, if the electric fields are small, the non-linear terms make a negligible contribution to the polarization, and hence can be neglected; this leads to the simple linear response. Clearly, when the electric fields are high, the nonlinear terms become relevant and the nonlinear effects come into play.

6.6.2 NLO Effects: Second Harmonic Generation

The optical field of an electromagnetic radiation with frequency, ω, varies in a symmetric sinusoidal form in time with frequency, ω. When it is incident on a material, it induces a polarization which varies with the same frequency, and gives rise to an optical field radiated by the material; this is the transmission of light through the material. If the process is linear, the field radiated is also symmetric, sinusoidal with the same frequency, ω.

In an NLO material, the nonlinear dependence of polarization on the incident field makes the polarization (and hence the radiated field) oscillate with frequency, ω, but in an unsymmetric form; the case with non-centrosymmetric media (Figure 6.17) is shown including time variation of the field and polarization, in Figure 6.18(a).

This polarization (and hence the radiated field) is a superposition of waves of various frequencies; mathematically, this is described as the Fourier decomposition of a periodic function in terms of various

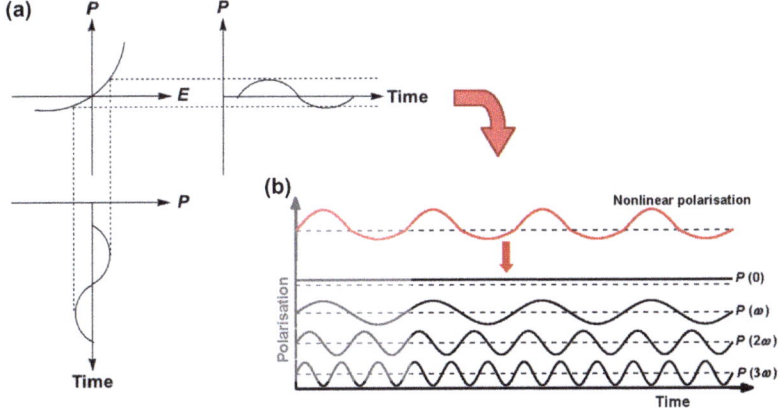

Figure 6.18 (a) Variation with time, of the incident radiation field (E) and the polarization (P) induced in a non-centrosymmetric NLO medium, and (b) Fourier decomposition of the nonlinear polarization.

harmonics. The phenomenon is represented pictorially in Figure 6.18(b).

- $P(\omega)$ is the fundamental polarization leading to radiation with the same frequency.
- The frequency independent term, $P(0)$ simply generates a voltage across the material, and this effect is called **optical rectification**.
- $P(2\omega)$ generates radiation with double the frequency as the fundamental; this is **second harmonic generation**. Higher harmonic generations occur in a similar way.

   Second harmonic generation (SHG) is a quadratic nonlinear optical effect. The incident field with frequency ω, can be represented as, $E(\omega) \propto \sin(\omega t)$. The quadratic response arises from the $E^2(\omega)$ term. $E^2(\omega) \propto \sin^2(\omega t)$. Since $\sin^2(\omega t) = 1 - \cos(2\omega t)$, it is clear that the second order polarization terms, $P^{(2)}(\omega) \propto [1 - \cos(2\omega t)]$. The first term which is frequency independent corresponds to optical rectification, and the second term with a 2ω dependence corresponds to SHG. Symmetry of the material has a critical role in determining the type of NLO effects that it can exhibit.
- In a medium (molecule or material) that is centrosymmetric (*i.e.*, possessing the center of inversion symmetry), inversion of the incident electric field will result in exact inversion of the polarization

   $E \rightarrow -E \Rightarrow P \rightarrow -P$. This can be expressed as: $P(-E) = -P(E)$.
- The equation for $P_i(\omega)$ in Section 6.6.1 shows that this happens only if the even order terms (with even powers of E) are strictly zero. The reasoning is as follows.
- Omitting the vector/tensor components and frequency dependence for simplicity, the equation can be written as:

$$P(E) = \chi^{(1)}E + \chi^{(2)}E^2 + \chi^{(3)}E^3 + \chi^{(4)}E^4 + \ldots$$

$$\therefore P(-E) = \chi^{(1)}(-E) + \chi^{(2)}(-E)^2 + \chi^{(3)}(-E)^3 + \chi^{(4)}(-E)^4 + \ldots$$

$$= -\chi^{(1)}E + \chi^{(2)}E^2 - \chi^{(3)}E^3 + \chi^{(4)}E^4 + \ldots$$

   Note that, $-P(E) = -\chi^{(1)}E - \chi^{(2)}E^2 - \chi^{(3)}E^3 - \chi^{(4)}E^4 + \ldots$.
- From the last two equations, it is clear that the equation, $P(-E) = -P(E)$ can be satisfied only if $\chi^{(2)}$, $\chi^{(4)}$, etc., vanish exactly. In other words, the quadratic NLO effects such as SHG cannot be observed in centrosymmetric media.

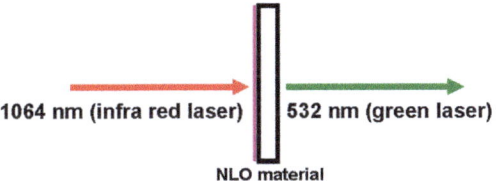

Figure 6.19 Illustration of the SHG process; doubling of the frequency corresponds to halving of the wavelength.

- Indeed, if the medium has no center of inversion, $E \to -E \Rightarrow P \to P'$. The above arguments are not applicable, and the system can exhibit any of the NLO effects, depending on the actual value of the non-linear susceptibilities, $\chi^{(n)}$.
- A typical example of SHG is shown schematically in Figure 6.19.

6.6.3 Materials for Quadratic NLO Effects

A wide range of materials are known, exhibiting efficient NLO effects including SHG. These include inorganic materials, molecular crystals, polymers, *etc.*

- Examples of well-known inorganic materials for SHG are potassium dihydrogen phosphate, KH_2PO_4 (KDP), potassium titanyl phosphate, $KTiOPO_4$ (KTP) and β-barium borate, β-BaB_2O_4 (BBO). KDP is commonly used to generat a 532 nm laser beam from the fundamental beam of an Nd:YAG laser with wavelength 1.064 μm; at room temperature it exists in the paraelectric phase and belongs to the non-centrosymmetric space group $I\bar{4}2d$.
- $LiNbO_3$ is a popular material used for electro-optic applications (Section 6.6.4).

Several molecular crystals based on organic molecules, coordination complexes, and organometallic systems have been developed for NLO applications like SHG. The equation for nonlinear polarization (Section 6.6.1) can be re-written for a molecule as, molecular polarization,

$$p_i(\omega) = \alpha_{ij}E_j(\omega) + \beta_{ijk}E_j(\omega)E_k(\omega) + \gamma_{ijkl}E_j(\omega)E_k(\omega)E_l(\omega) + \dots$$

α = molecular polarizability, β = first hyperpolarizability, γ = second hyperpolarizability.

- β is the important characteristic at the molecular level, required to develop SHG materials. Further, proper orientation of the

β tensor components that add up to a high $\chi^{(2)}$ value of the crystal or bulk material, is required to obtain strong SHG.

- Even if β at the molecular level is high, if the crystal is centrosymmetric, the resultant $\chi^{(2)}$ will be zero, and SHG absent; non-centrosymmetric crystals on the other hand will have a non-vanishing $\chi^{(2)}$. A simple approach to understand how the β tensor components contribute vectorially to the $\chi^{(2)}$ is provided by the **oriented gas model.**[3]

$$\chi_{IJK}^{(2)} = Nf_I(2\omega)f_J(\omega)f_K(\omega) \sum_{ijk} \sum_{s=1}^{m} \left\{ \cos\theta_{I,i(s)}\cos\theta_{J,j(s)}\cos\theta_{K,k(s)} \right\} \beta_{ijk}$$

N is the number of unit cells per unit volume of the crystal, and $f_X(\omega)$ is the Lorentz local-field factor, $[(n_X^\omega)^2 + 2]/3$, n_X^ω being the refractive index. The $\cos\theta$ product represents the rotation of the molecular reference frame (i, j, k) to the crystallographic reference frame (I, J, K), and m is the number of molecules in the unit cell.

- A classic example of a simple molecular crystal that shows SHG is urea. Molecular assembly in the urea crystal mediated by extensive H-bonding is shown in Figure 6.20; it is a beautiful example of intermolecular interactions leading to a non-centrosymmetric lattice with bulk $\chi^{(2)}$ that enables an SHG response.
- Urea is often used as a standard; its SHG response is approximately three times that of KDP. Examples of some well-known molecules, crystals of which exhibit SHG are collected in Table 6.1; SHG responses of the microcrystalline samples relative to that of urea are also indicated. Chirality is often used (*e.g.*, MAP and NPP) as a tool to ensure the formation of a non-centrosymmetric crystal lattice.

An important consideration for efficient NLO responses is **phase-matching**. When waves of different frequencies and hence refractive indices of the material are involved, there can be a mismatch between the phases of such waves, which in many instances affect the efficiency of the NLO process. In the case of SHG, phase-matching condition is $k_2 = 2k_1$ where k_1 and k_2 are the wave vectors of the fundamental and second harmonic radiations. This can be rewritten in terms of the respective refractive indices, $n(\omega) = n(2\omega)$.

The non-centrosymmetric structure required for quadratic NLO effects, can be obtained from orientationally disordered and hence centrosymmetric materials, by applying strong electric fields to order the dipolar structures in the medium. This process is called '**poling**',

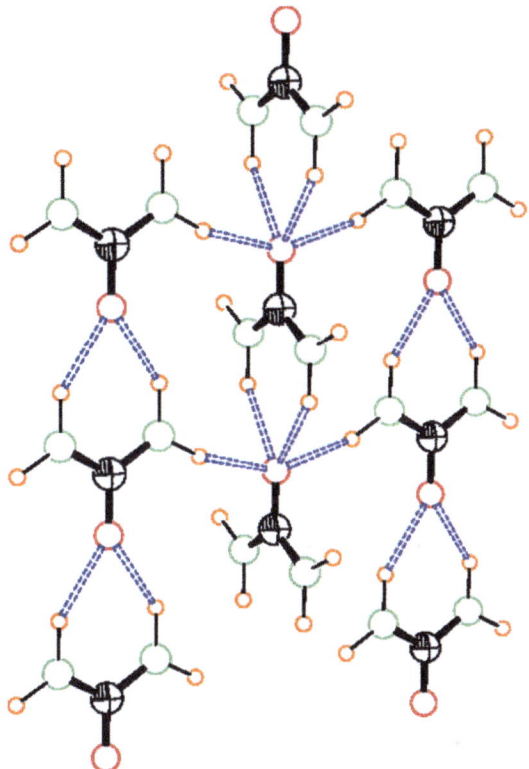

Figure 6.20 Molecular assembly in urea crystal, mediated by H-bonds (blue broken lines); C (black), O (red), N (green) and H (orange) atoms are indicated.

and effectively used in polymeric materials. The polymers may either have dipolar functional groups or dopants. An electric field is applied on the thin film material heated close to its glass transition temperature, and cooled with the poling field on. Orientational ordering of the dipoles breaks the symmetry. In many instances, cross-linking of the polymer in the poled state is used to stabilize the non-centrosymmetric structure against de-poling.

6.6.4 Electro-optic Effects

Electro-optic (EO) effects are the changes in the optical properties of materials in response to an electric field which varies slowly compared to the frequency of light. One set of EO effects relate to changes in the optical absorption upon application of a field, effects such as electroabsorption (change in absorption coefficient) and electrochromism (change in absorption wavelength). Another set of

Table 6.1 Examples of molecules, crystals of which exhibit SHG (standard abbreviations of the names are shown), and the SHG response of the microcrystalline samples relative to urea.

Molecule		SHG/U
MAP		10
mNA		40
MNA		80
DAN		115
NPP		150

important EO effects relate to changes in the refractive index upon application of an electric field.

- **Pockel's effect** is the linear EO effect in which the refractive index changes linearly with the applied field. This occurs in quadratic NLO materials; the medium causes a coupling between the electric field and the optical field. The nonlinear refractive index, $n = n_0 + \Delta n$, where the change in refractive index, $\Delta n \propto \chi^{(2)} E$. This effect is used in electro-optic modulators.
- A Mach–Zehnder interferometer can be used to achieve amplitude modulation (Figure 6.21). A beam splitter divides a continuous wave (CW) optical signal into two paths at **A**. If a GHz frequency electric field is imposed on one path with an EO material, the phase (ϕ) of the beam on this path is modulated at the same frequency. This determines the interference when the beams from the two paths meet at **B**, resulting in the output signal amplitude being modulated. Electrical signals can be mapped on to photonic signals in this way.
- The **Kerr effect** is the quadratic EO effect in which the refractive index changes quadratically with the applied field. The refractive

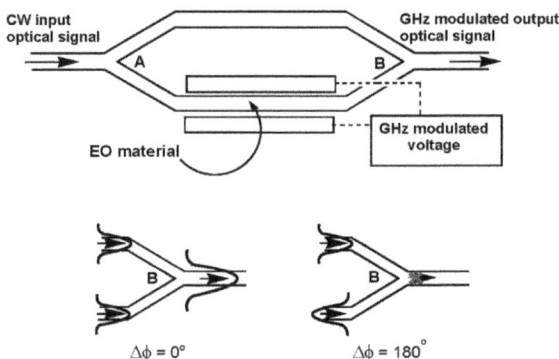

Figure 6.21 Schematic representation of the amplitude modulation of a photonic signal using a Mach–Zehnder modulator; change in phase, $\Delta\phi$ between the beams arriving at **B** leading to constructive/destructive interference are shown.

index change is proportional to the square of the electric field ($\Delta n \propto E^2$). The change can also be expressed as being proportional to the intensity of the light; $n = n_0 + n_2 I$. n_2, the second order nonlinear refractive index is related to $\chi^{(3)}$. Being a third order nonlinear effect, the quadratic EO process can be observed in centrosymmetric as well as non-centrosymmetric media.

- A Kerr gate functions by using the Kerr effect to open/close the transmission of a photonic signal. The principle is illustrated in Figure 6.22. The polarized beam transmitted through an EO ($\chi^{(3)}$) crystal retains the polarization and hence is stopped by the cross polarizer. However, it undergoes a polarization rotation when the refractive index of the EO crystal is modified by another laser beam, and hence is not fully cut off by the cross polarizer; laser 2 acts as a photonic signal to open the gate for laser 1. Photons controlling photons in a Kerr gate can be considered as the photonic equivalent of a transistor based gate in electronics.

6.6.5 Third Harmonic Generation and Organic Polymers

Like the intensity dependent refractive index (Kerr effect), the third harmonic generation (THG) also results from the third order nonlinear susceptibility, $\chi^{(3)}$ at the bulk material level, or the second hyperpolarizability γ at the molecular level.

As shown in Section 6.6.2 for SHG, generation of the third harmonic can be understood as follows. $E^3(\omega) \propto \sin^3(\omega t)$.

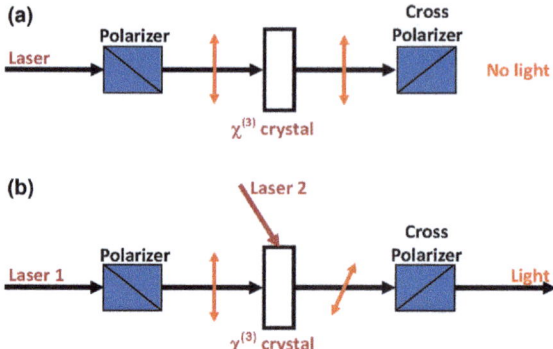

Figure 6.22   Working principle of a Kerr gate: (a) absence of polarization change and the crossed polarizer acting as a 'close' gate; (b) the change of polarization effected by laser 2 causing the gate to 'open'.

Since $\sin^3(\omega t) = \dfrac{1}{4}[3\sin(\omega t) - \sin(3\omega t)]$, it is clear that the third order polarization terms, $P^{(3)}(\omega) \propto [3\sin(\omega t) - \sin(3\omega t)]$.

- The ω term relates to the optical Kerr effect discussed in the previous section.
- The 3ω term gives rise to the output at triple the frequency of the fundamental; this is the process of THG.

Due to the relatively low values of $\chi^{(3)}$, the third order effects are generally very weak. Conjugated polymers are some of the most well-known materials developed for efficient THG. The γ values are known to increase with the conjugation length, typically up to molecular lengths of about 2 nm. Polydiacetylene and poly(*p*-phenylenevinylene) (PPV) with various substituents are examples of conjugated polymers that have been extensively studied as materials for THG applications.

References

1. C. Kittel, *Introduction to Solid State Physics*, Wiley, New York, 5th edn, 1976.
2. R. W. Gurney and N. F. Mott, *Proc. R. Soc. Lond. A*, 1938, **164**, 151.
3. J. Zyss and J. L. Oudar, *Phys. Rev. A*, 1982, **26**, 2028.

7 Synthesis and Fabrication

7.1 Synthesis

Synthesis is the starting point for any real world investigation and application of molecules or materials. The structural and functional aspects discussed in the various earlier chapters are in some sense, simultaneously the motivation as well as the logical follow up of the synthesis and fabrication of materials.

7.1.1 Organic and Inorganic Compounds

Organic molecules are synthesized from precursor molecules by the breaking and making of covalent bonds. The power of organic synthesis has grown exponentially through improved understanding of mechanistic pathways involved in the processes. An enormous repertoire of synthetic methodologies has been developed over time. A few simple examples of organic synthesis are collected in Figure 7.1; consult the many standard textbooks on organic synthesis for more case studies.

Macromolecules form an important class of materials; they are formed through extended covalent bond formation that connects hundreds or thousands of monomer molecules into the polymer structure.

Many inorganic compounds are prepared through simple double decomposition reactions involving ionic species.

- A simple example is: $AgNO_3$ (aq) $+ NaCl$ (aq) $\rightarrow AgCl$ (s) $+ NaNO_3$ (aq). Recognizing the dissolved state and precipitation of the

Core Concepts for a Course on Materials Chemistry
By T. P. Radhakrishnan
© T. P. Radhakrishnan 2023
Published by the Royal Society of Chemistry, www.rsc.org

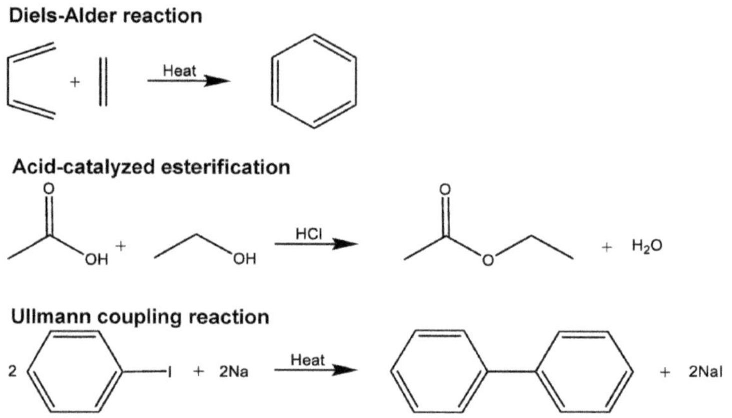

Diels-Alder reaction

Acid-catalyzed esterification

Ullmann coupling reaction

Figure 7.1 Some selected organic reactions.

different species involved, it would be appropriate to write this reaction as:

Ag^+ (aq) $+ NO_3^-$ (aq) $+ Na^+$ (aq) $+ Cl^-$ (aq) $\rightarrow AgCl$ (s)$\downarrow + Na^+$ (aq) $+ NO_3^-$ (aq). Strictly, these are all equilibria with each species present at a specific concentration.

• Acid–base reaction is another common example:

$$HCl \ (aq) + NaOH \ (aq) \rightarrow NaCl \ (aq) + H_2O \ (l).$$

Metal complexes are formed through the coordinative attachment of ligands to metal ions: $[Ni(H_2O)_6]Cl_2$ (aq) $+ 2NaOH$ (aq) $\rightarrow [Ni(H_2O)_4$-$(OH)_2]$ (aq) $+ 2H_2O$ (l) $+ 2NaCl$ (aq). Extended formation of coordination bonds leads to coordination polymers; specialized structures such as metal organic frameworks are examples of such materials.

Assembly of molecules leads to molecular materials. These could be in the form of crystals, nanocrystals, liquid crystals, colloids, thin or ultrathin films, amorphous solids, gels, *etc.* Many of the inorganic materials form solids through extended bonding interactions. Sodium chloride is a prototypical example. The concept of a single independent unit, which can be called a molecule, is not very clear or relevant in these materials. The general concept of synthesis of such materials is shown schematically in Route A of Figure 7.2, and contrasted with the case of molecular materials obtained through the two-step Route B.

Some of the basic concepts involved in the various materials fabrication strategies are discussed in the major sections later. In this section, we consider some special instances of solid state reactions.

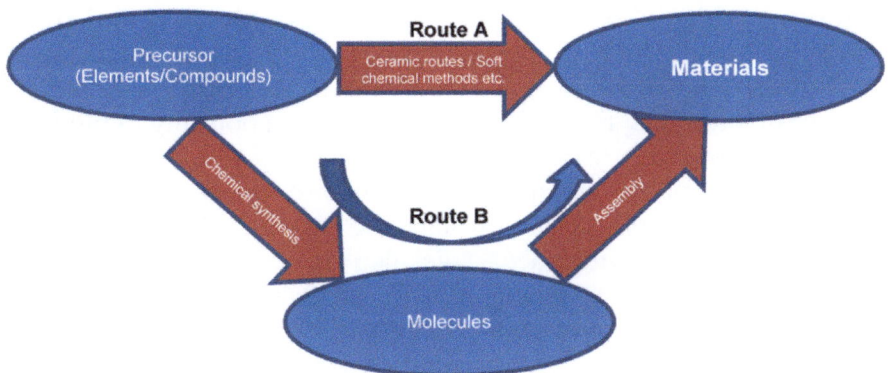

Figure 7.2 Two-step synthesis/fabrication of molecular materials (Route B) contrasted with the generally direct route to other types of materials (Route A).

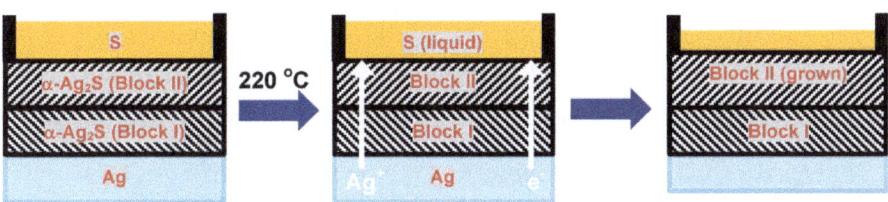

Figure 7.3 Wagner's experiment demonstrating the diffusion of Ag^+ ions through α-Ag_2S followed by electrons, leading to the formation of Ag_2S at the interface of Block II and molten S. Note the depletion of the Ag block and S liquid, and growth of the Ag_2S (Block II), Ag_2S (Block I) remaining unchanged.

7.1.2 Solid State Reactions

Reactions in the solid state are critical in various processes including heterogeneous catalysis, corrosion, solid state photochemistry, crystal-to-crystal transformations, *etc.* Catalytic processes are generally associated with high rates; on the other hand, those such as corrosion are usually slow. More details about some of the topics in this section can be found in ref. 1.

Diffusion of atoms or ions in the solid state plays a basic role in solid state reactions, specifically, limiting the reaction rates.

- **Wagner**'s experiment on Ag_2S formation from Ag and S demonstrated beautifully the importance of cation diffusion in solid state reactions. The general idea is shown schematically in Figure 7.3.
- α-Ag_2S is an ionic (Ag^+) as well as an electronic conductor.

- A block of Ag is separated from S using two blocks (I and II) of α-Ag$_2$S. When the system is heated at 220 °C, the S melts. It is observed that Block II grows at the interface with S, advancing into the molten S; the Ag block shows a corresponding depletion. Block I remains unchanged.
- The mechanism that explains this observation is that Ag$^+$ ions from the Ag block diffuse through the two α-Ag$_2$S blocks, reaching the α-Ag$_2$S/S interface. This also drives a parallel flow of electrons from the Ag block to the interface, that reduces S to S^{2-}. The Ag$_2$S formed, grows at the interface, advancing into the molten S. Note that S^{2-} does not diffuse in the reverse direction. Ionic radii of Ag$^+$ and S^{2-} are 115 pm and 184 pm respectively; the α-Ag$_2$S lattice allows diffusion of Ag$^+$ but not S^{2-}.
- A similar mechanism is invoked to explain the formation of spinels and silicates. For example, MgAl$_2$O$_4$ formation from MgO and Al$_2$O$_3$ occurs by diffusion of Mg^{2+} and Al^{3+} ions in opposite directions, O^{2-} remaining immobile. The reactions at the two interfaces may be written as follows:

$$\text{MgO/MgAl}_2\text{O}_4 \text{ interface: } 4\text{MgO} + 2\text{Al}^{3+} - 3\text{Mg}^{2+} \rightarrow \text{MgAl}_2\text{O}_4$$

$$\text{MgAl}_2\text{O}_4/\text{Al}_2\text{O}_3 \text{ interface: } 4\text{Al}_2\text{O}_3 + 3\text{Mg}^{2+} - 2\text{Al}^{3+} \rightarrow 3\text{MgAl}_2\text{O}_4.$$

Corrosion of metals (oxidation starting on the surface) is a technologically important solid state reaction. Ionic conductivity of the metal oxide film formed on the surface of metals plays a crucial role in this process.

- The schematic representation in Figure 7.4 shows that O$_2$ gas from the environment can diffuse through the pores and cracks in the

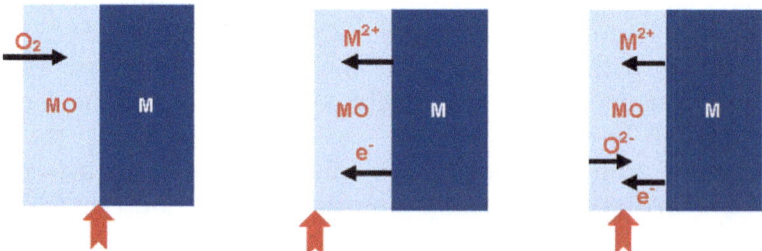

Figure 7.4   Diffusion of O$_2$ through the cracks/pores, and conduction of M^{2+} and O^{2-} through the metal oxide (MO) layer lead to the corrosion (oxidation) of metal (M). The position where the MO formation occurs depends on the relative speeds of the various species involved and is indicated by the upward pointing red arrows.

metal oxide (MO) film. Metal ions, M^{2+} are conducted through the MO film. Electrons from the metal reduce the O_2 molecules; the O^{2-} ions formed combine with the M^{2+} ions forming MO. The position of MO growth (indicated by the vertical red arrow) depends on the relative speed of O_2 diffusion and conduction of M^{2+} as shown in Figure 7.4. When they are comparable, the MO growth happens in the middle region of the MO film as shown.

- The free energy change (ΔG) that drives the MO formation is dominated by the enthalpic term, as the entropy change would be negative due to the localization of $O_2(g)$ when forming MO(s).
- The MO formation can be reduced by limiting the ionic conductivity. A typical example is the case of NiO in which cation vacancies lead to high ionic conductivity. Addition of aliovalent impurities like Li_2O (Section 2.3.2) fills these vacancies (as two Li^+ replace one Ni^{2+}), decreases the ionic conductivity, and reduces the oxidation.
- The rate of oxide formation (x = thickness of the oxide layer, t = time) is proportional to the flux of electrons, M^{2+}, O^{2-} *etc.* This can be expressed as: $\dfrac{dx}{dt} \propto J$. Since the flux is inversely related to the thickness, $\dfrac{dx}{dt} \propto \dfrac{1}{x}$, integration gives:

$$\int_0^x x\,dx \propto \int_0^t dt \Rightarrow x^2 \propto t \text{ or } x \propto \sqrt{t} \text{ } i.e. \text{ parabolic growth behavior.}$$

Roasting and calcination are typical processes by which metal oxides are formed on heating the corresponding metal ores. The former involves heating the solid at a temperature below the melting point, in presence of air (*e.g.*, metal sulfide is converted to metal oxide); the latter is usually a thermal decomposition (*e.g.*, metal carbonate is converted to the metal oxide).

An illustrative example of the '**ceramic method**' for solid state synthesis, sometimes called the 'shake-and-bake' approach, is the preparation of the well-known 'high T_C superconductor', $YBa_2Cu_3O_{7-x}$ (Section 2.4.2 and Section 4.4.4). The important role of phase diagrams can be illustrated by considering the formation of this material. The ternary phase diagram with the three oxides, Y_2O_3, BaO, and CuO is shown in Figure 7.5;[2] it corresponds to the phases present at 950 °C, in air.

- The three vertices represent the individual oxides (purple filled circles and labels); in the presence of CO_2, BaO exists as $BaCO_3$.
- Each edge of the triangle represents the binary compositions; some specific stable compositions are indicated (red filled circles and labels).

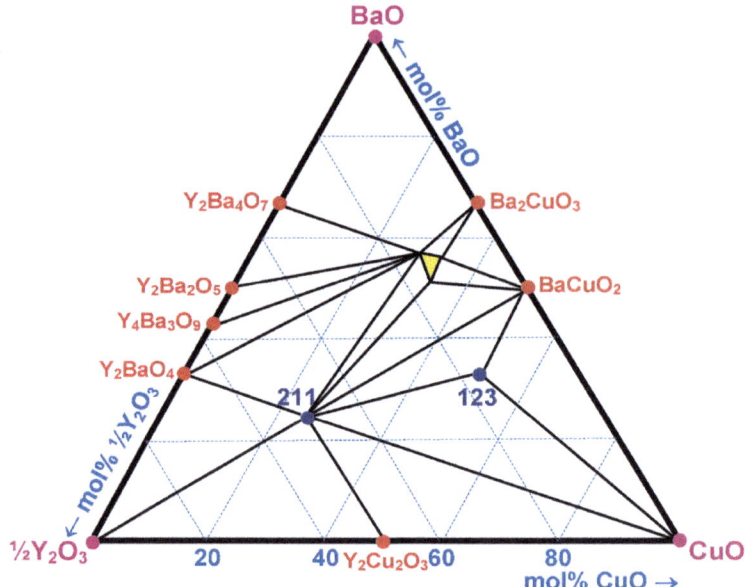

Figure 7.5   Phase diagram of the $\frac{1}{2}Y_2O_3$ – BaO (BaCO$_3$) – CuO system at 950 °C, in air. Drawn schematically based on Figure 1 in ref. 2. See ref. 2 for more details.

- Several ternary systems can be identified in the diagram; the specific cases of **211** (Y$_2$BaCuO$_5$) and **123** (YBa$_2$Cu$_3$O$_{7-x}$) are indicated (blue filled circles and labels).
- The yellow shaded region is a typical solid solution of some ternary systems.
- The phase diagram allows the identification of stable phases and composition of precursor materials, as well as the temperature/ pressure and environmental conditions needed to synthesize the desired unique material.

Other interesting solid state reactions include photochemical processes such as dimerizations and polymerizations in crystals, and crystal-to-crystal transformations.

- The pioneering work on photochemical solid state dimerization of cinnamic acid by Schmidt and coworkers in the 1960s established the critical role of topochemical control; reactions in the solid state proceed with minimum movement of atoms and molecules. An early review of photochemical reactions in organic crystals can be found in ref. 3.

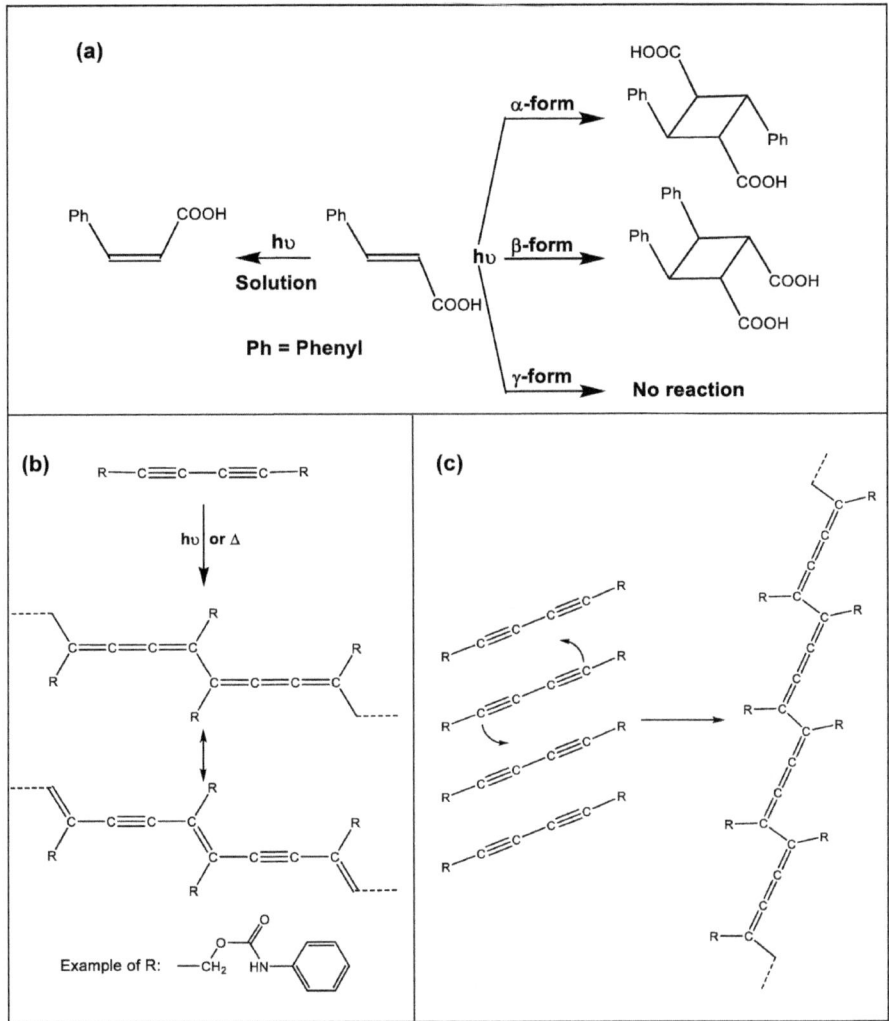

Figure 7.6 (a) Topochemical photodimerization of *trans*-cinnamic acid in different crystalline polymorphic forms (α, β and γ) contrasted with the photochemical reaction in solution. (b) Polymerization of diacetylene to poly(diacetylene); the two resonance structures of the latter are shown. (c) Schematic diagram showing the topochemical nature of the polymerization process involving cooperative shearing of the ladder structure in the single crystal.

- The illustrative example in Figure 7.6(a) shows the relevance of the distance between the molecules and their relative orientation. Photoirradiation of *trans*-cinnamic acid in solution produces *cis*-cinnamic acid. However, in the solid state, dimerization products are obtained depending on the specific crystal

polymorph used; the α-form with a separation of 3.6–4.1 Å between the double bonds and a centrosymmetric orientation of
the adjacent molecules gives α-truxillic acid, whereas the β-form
with a separation of 3.9–4.1 Å between the double bonds and a
translation relation between adjacent molecules produces
β-truxillic acid. The γ-form with a larger separation of 4.7–5.1 Å
between the double bonds and a translation relation between
adjacent molecules, does not yield any product.

- The logical extension to **photopolymerization** is also well known.
  A famous example is the formation of poly(diacetylene) from
  diacetylene by heat or UV irradiation.[4] Semiconducting poly-
  (diacetylene) has the butatriene or the alternating double-triple
  bond structures (Figure 7.6(b)). The topochemical reaction is
  enabled by the short (~3 Å) distance between the triple bonds of
  adjacent molecules in the ladder-type organization of diacetylene
  molecules in the crystal; the polymerization involves the tilting of
  the rungs or cooperative shearing of the ladder (Figure 7.6(c)).
- Several examples of **crystal-to-crystal transformation** are known,
  wherein a single crystal transforms to another with or without
  chemical reaction happening in the crystalline state. Some are even
  known to be reversible. A more general concept is that of a **topotactic
  reaction**. This is a solid state process involving the displacement or
  exchange of atoms; the crystalline orientation of the product is
  determined by that of the reactant. It may be accompanied by the
  loss of byproducts or incorporation of additional reactants.

7.2 Crystallization

Crystals play a vital role in materials chemistry and materials science
in general, since they not only provide access to unique properties
and functions, but also allow rigorous determination of the organization of atoms, ions or molecules (Chapter 1) providing critical
insight into the nature of the interactions that bind them together to
form the material. The crystallization process and crystals are useful
in a wide variety of scenarios:

- Crystallization is often an effective method for purifying a material.
- Single crystals have unique uses and advantages in some
  specialized device applications, *e.g.,* Si in solar cells.
- Crystallization allows resolution of racemic mixtures and abso-
  lute structure determination (recall the famous experiment by
  Louis Pasteur).

- In specialized investigations including spectroscopy, single crystals provide far more detailed information than micro-crystalline powder samples.
- Bio-crystallization has significant medical implications, *e.g.,* osteoporosis and kidney stone formation.
- Polymorphism (a material forming different crystal structures) is very important in the pharmaceutical industry, with a significant bearing on intellectual property rights. The polymorph solubility also determines the drug availability (in solution), and can be exploited for controlled release.
- Properties of macromolecular materials are strongly dependent on the extent of their crystallinity.

The process of formation of crystals, **crystallization**, can in general occur from the melts of the solid, or solutions in suitable solvents; crystals can also be formed from the gaseous state directly. The basic events involved are **nucleation** and **growth**. Fundamental thermo-dynamic aspects of nucleation are discussed in Sections 7.2.1 to 7.2.4. Formation of nuclei from the molten state (1-component system) on cooling is discussed; the same basic concepts apply to the formation of nuclei from solution (2-component system) upon increasing the concentration above saturation limits.

The most common approach for crystal growth in chemistry is from solution. Some general approaches are discussed in Section 7.3. Other crystal growth techniques from the gaseous and melt states as well as specialized aspects of crystal growth and purification in the solid state are presented in Section 7.4.

7.2.1 Thermodynamic Aspects

A single component system involved in crystal growth from the melt, is the simplest situation to consider. This is essentially a liquid–solid phase transition. The basic process of freezing of a liquid (melt) to form the solid can be visualized using the plot of chemical potential (or free energy) as a function of temperature (Figure 7.7).

- At the normal melting (or freezing) temperature, T_M, the solid (s) and liquid (l) are in equilibrium, s⇌l; their chemical potentials are equal.
- Below and above the T_M, s and l respectively are more stable. On cooling the liquid below T_M, the solid should form; however, this transformation is often hindered by kinetic factors. Hence a

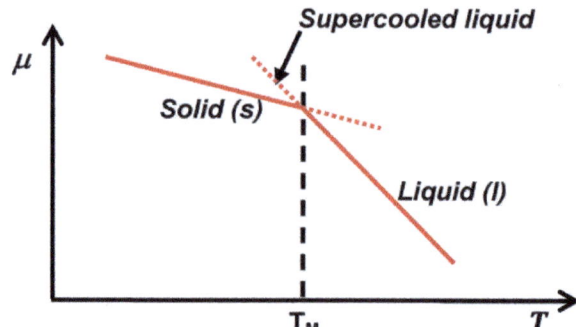

Figure 7.7   Schematic plot of chemical potential (μ) *versus* temperature (T) across the normal melting temperature (T_M) at which the solid and liquid are in equilibrium. The metastable supercooled liquid state is indicated.

supercooled state can exist where the liquid remains in a thermodynamically metastable state below the normal T_M (similarly, at the liquid–vapor transition, the liquid can stay in a superheated state above the boiling point).

- When small particles of s form, the high surface free energy will dominate over the bulk free energy causing the particle to revert back to the l state. If the particle of s somehow manages to reach a critical size, r^*, the bulk free energy overwhelms the surface free energy and a stable particle forms; this is **nucleation**.
- These thermodynamic considerations are very general, and apply to several phase transitions.

7.2.2   Homogeneous Nucleation

When a particle of solid forms from the liquid melt, two free energies come into play: the free energy due to bonding interactions in the bulk of the particle (a stabilizing factor, called the volume free energy), and the surface free energy due to the unsatisfied bonding interactions of the atoms or molecules on the particle surface (a destabilizing factor).

Balance between these two factors decides whether a nucleation occurs or not. When the nucleation occurs in a homogeneous medium (the liquid in this case) with no interfaces or surfaces that create heterogeneity, it is called homogeneous nucleation. The basic concept of the classical theory of nucleation is as follows. Consider the processes under constant temperature and volume conditions, and the Helmholtz free energy changes for the liquid → solid transition.

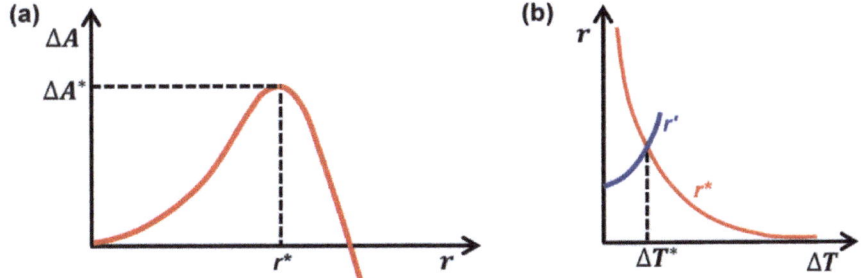

Figure 7.8 (a) Variation of the free energy change with radius of the solid particle forming in the melt. The critical radius (r^*) corresponding to the plot maximum (ΔA^*) is indicated. (b) Variation of the critical radius, r^* (red line) and the radius of the largest probable nucleus, r' (blue line) with the extent of supercooling, ΔT. The supercooling required for homogeneous nucleation to occur (ΔT^*) is shown.

- Surface free energy change: $\Delta A_S = \gamma \Delta \sigma = 4\pi r^2 \gamma$ ($\gamma =$ surface tension of the new interface created, solid–liquid in this case, $\Delta \sigma =$ change in surface area of the particle formed, $r =$ radius of the particle assumed to be spherical).
- Volume free energy change: $\Delta A_V = -\dfrac{4}{3}\pi r^3 |\Delta A_v|$ ($\Delta A_v = \Delta A$ per unit volume).
- The total free energy change, $\Delta A = \Delta A_S + \Delta A_V = 4\pi r^2 \gamma - \dfrac{4}{3}\pi r^3 |\Delta A_v|$. The two opposing (sign) dependencies lead to a maximum in the plot of ΔA versus r (Figure 7.8(a)); the value of r at the maximum in the free energy profile is the **critical radius, r^***.
- The critical radius can be determined by finding the maximum of the curve:

$$\left[\frac{\partial(\Delta A)}{\partial r}\right]_{r=r^*} = 0 = 8\pi r^* \gamma - 4\pi (r^*)^2 |\Delta A_v| \Rightarrow r^* = \frac{2\gamma}{|\Delta A_v|}.$$

- For $r < r^*$, increase of r causes the free energy to increase and any solid particle that is formed, to disintegrate back to the liquid form. When $r > r^*$, the decreasing free energy enables the nucleus formed, to grow, *i.e.*, crystal growth is initiated.

In the equation for r^*, γ is essentially independent of temperature, but ΔA_v is not. Hence the critical radius has a temperature dependence (Figure 7.8(b)). This can be shown as follows.

- The Helmholtz free energy change per unit volume can be related to the changes in internal energy (E) and entropy (S) associated with

the freezing, the liquid→solid phase transition, $\Delta A_v = (E_s - TS_s)$
$-(E_l - TS_l) = (E_s - E_l) - T(S_s - S_l) = \Delta E_{l \to s} - T\Delta S_{l \to s}$.

- Assuming that $\Delta E \cong \Delta H$, and ΔH and ΔS are nearly independent of temperature near the T_M, and since the entropy change for freezing is the enthalpy change divided by the freezing temperature,

$$\Delta A_v = \Delta H_{l \to s} - T\Delta S_{l \to s} = \Delta H_{l \to s} - T\frac{\Delta H_{l \to s}}{T_M} = \frac{\Delta H_{l \to s} \cdot \Delta T}{T_M}, \text{ where}$$

$\Delta T = T_M - T$, the extent of supercooling.

- The critical radius can now be expressed as: $r^* = \dfrac{2\gamma T_M}{|\Delta H_{l \to s} \cdot \Delta T|}$.

The plot in Figure 7.8(b) shows this hyperbolic dependence of critical radius on the supercooling.

Depending on the extent of supercooling (ΔT), size of the largest probable nucleus (r' in Figure 7.8(b)) will increase. The extent of supercooling to reach $r' \geq r^*$ is the supercooling required for homogeneous nucleation to occur (ΔT^*). When $T \lesssim T_M$, or $\Delta T \sim 0$, r^* is very large and r' very small (Figure 7.8(b)). So nucleation does not occur. When ΔT increases, r' increases and r^* decreases, and eventually nucleation occurs.

An important prediction of the nucleation theory, usually called the classical nucleation theory, is the rate of nucleation, R.

- R is determined primarily by two factors, the number of nuclei at the top of the barrier (shown in Figure 7.8(a)) and the probability that the nucleus formed will grow and not dissolve back.
- The first factor is determined by the number of nucleation sites, N and the Boltzmann factor that gives the population of nuclei at the top of the barrier, $e^{-\frac{\Delta A^*}{k_B T}}$ where k_B is the Boltzmann constant and T is the temperature.
- The second factor is controlled by the rate at which the atoms or molecules attach to the nucleus, J and the probability that the nucleus moves to the right in the plot in Figure 7.8(a); the latter is known as the Zeldovich factor, Z. The term ZJ is proportional to the diffusion coefficient; hence it is proportional to the temperature, T and inversely proportional to the coefficient of viscosity.
- Overall, the rate of nucleation can be expressed as: $R = NZJe^{-\frac{\Delta A^*}{k_B T}}$

7.2.3 Heterogeneous Nucleation

The discussion on homogeneous nucleation assumed the nucleus to be spherical. In a homogeneous medium, this is generally likely. On a substrate where the liquid can spread, the smaller the contact

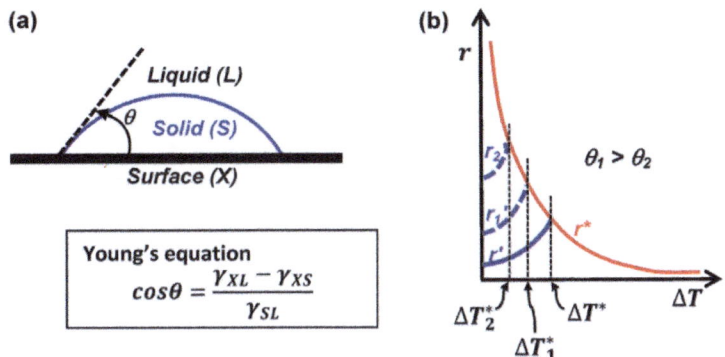

Figure 7.9 (a) Nucleus formed on a surface; the contact angle, θ is indicated; Young's equation relates the contact angle with the various interfacial tensions, γ. (b) Radius of the largest probable nucleus for homogeneous nucleation r' (blue line) compared to that for heterogeneous nucleation (r'_1, r'_2) with different contact angles (θ_1, θ_2); intersection with the plot of the critical radius, r^* (red line) shows the decreasing supercooling ($\Delta T^* > \Delta T^*_1 > \Delta T^*_2$) required for nucleation.

angle (Figure 7.9(a)), the larger the effective radius of the nucleus (solid) being formed, and hence the easier it is to exceed the critical radius. Hence the supercooling required is effectively less, facilitating easier nucleation and crystal growth. This is the physical basis for heterogeneous nucleation.

- The contact angle is determined by the various interfacial (surface) tensions, as per the Young's equation (Figure 7.9(a)).
- The comparison between the supercooling ΔT required for homogeneous nucleation with that for heterogeneous nucleation with two different contact angles ($\theta_1 > \theta_2$) is shown in Figure 7.9(b). In the case of very low contact angle (high spreading), the nucleation occurs close to T_M (*i.e.,* $\Delta T \sim 0$).

7.2.4 Two-step Nucleation

Experimental studies of nucleation have often indicated significant deviation from the predictions of the classical nucleation theory, for example, in terms of the rate of nucleation. One of the factors pointed out is that the process does not occur in a single step through a single barrier as shown in Figure 7.8(a).

The two-step nucleation concept[5] is that the atoms or molecules first come together in a disordered cluster (increasing the local

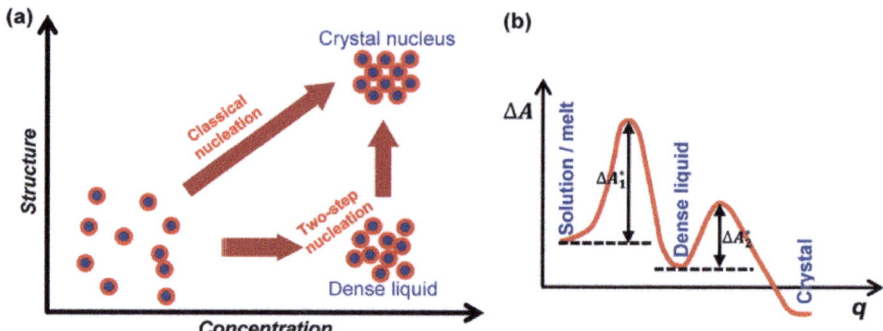

Figure 7.10 (a) Schematic diagram showing the two order parameters 'concentration' and 'structure' to represent the classical nucleation and two-step nucleation processes. (b) Variation of the free energy change (ΔA) with the nucleation coordinate (q) for the two-step process; the free energy maxima for the two steps are indicated.

concentration), and subsequently organize into the ordered crystal lattice structure.

- These can be visualized in terms of two order parameters concentration and structure; as shown in Figure 7.10(a), the classical nucleation goes from individual components in one step to the final ordered crystal, whereas the two-step nucleation goes through the intermediate stage called the 'dense liquid', which is essentially a disordered aggregate.
- The corresponding free energy profile (Figure 7.10(b)) shows two maxima with the intermediate 'dense liquid' structure.
- Crystallization of several macromolecules and even some small molecules has been shown to follow the two-step nucleation process.

7.3   Crystal Growth from Solution

Many of the concepts presented in this and the following sections are based on the excellent review of crystal growth techniques by Hulliger.[6]

7.3.1   Phase Diagram

The 2-component phase diagram (solute concentration–temperature, under constant pressure conditions), provides the thermodynamic background for understanding the crystallization from solution.

- The solute concentration (as mol fraction, x) is plotted on the x-axis and the temperature (T) on the y-axis in Figure 7.11.
- The phase boundaries (also called the **liquidus**) between the solution (1 phase), frozen solvent + solution (2 phases) and crystallized solute + solution (2 phases), are also associated with the superheating and supercooling regimes (Section 7.2.1).
- The phase boundary between the 2-phase regions noted above, with the mixture of the two solids (2 phases) is called the **solidus.**
- The eutectic point is defined by a fixed concentration and temperature (zero degrees of freedom); the solution freezes directly into the solid mixture without a liquid + solid phase. Example: NaCl + water (common salt solution) has a eutectic point at 23.3 weight% of NaCl ($x = 0.086$) and -21.1 °C; the phase diagram is more complex than that shown in Figure 7.11, at higher concentrations.

The ΔT–Δx region shown in Figure 7.11 is suitable for crystal growth from solution. The boundaries are defined by several complex parameters including the cooling speed, hydrodynamic factors, nature of solvation, presence of impurities, *etc.*

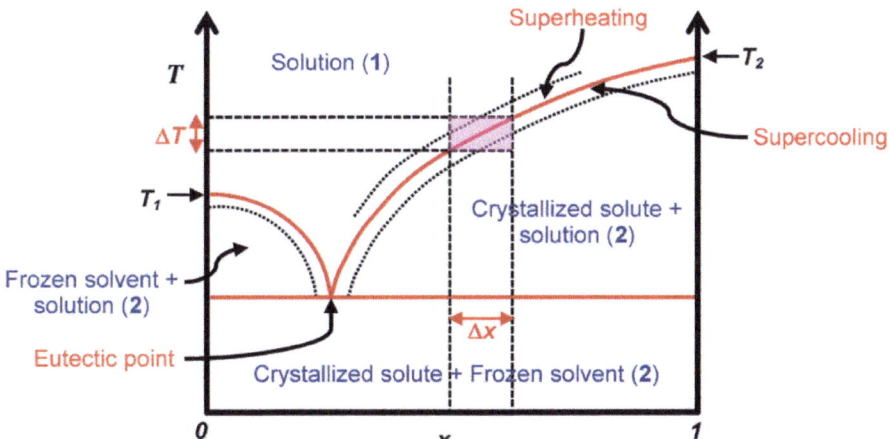

Figure 7.11 Concentration–temperature phase diagram (at constant pressure) of a solution; full red lines represent the phase boundaries, and broken black lines, the limits of supercooling and superheating of the solution. The number of phases in each region is shown in parenthesis. The ideal crystallization regime (pink) in terms of concentration (Δx) and temperature (ΔT) is indicated. The eutectic point corresponds to the direct freezing of the solution to the solid mixture, melting point of the solvent (T_1) and the solute (T_2) are marked. Drawn schematically based on Figure 7 and 20 in ref. 6.

7.3.2 Isothermal Methods

The **evaporation** method is perhaps the most common and simple procedure for growing crystals from solution.

- Typically, a solution with concentration slightly below the saturation level is maintained at a constant temperature that is well below the boiling point of the solvent. Upon slow evaporation of the solvent, the solute concentration increases and the crystals emerge, after the concentration exceeds the solubility limit at the relevant temperature and nucleation sets in.
- Controlled nucleation or introduction of a seed crystal can enable efficient single crystal growth. Removal of initial crude nuclei by redissolution can be helpful.
- The rate of evaporation can be regulated by controlling the vents (Figure 7.12(a)).
- For improved crystal growth, those formed are removed carefully. In more elaborate systems, the concentration levels can be maintained by condensing back the evaporated solvent and the saturation level maintained by a supply of the nutrient (solute) as shown in the apparatus in Figure 7.12(b).

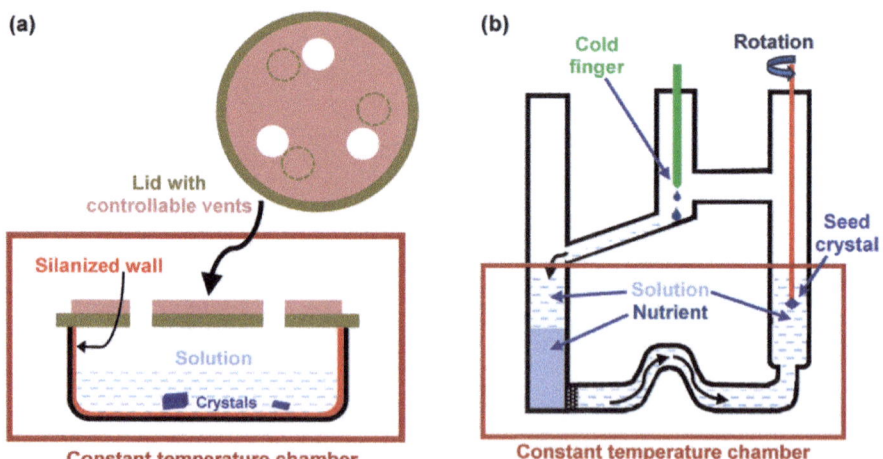

Figure 7.12   Crystal growth by isothermal evaporation: (a) A simple setup with controllable vents that regulate the evaporation rate; the silanized wall inhibits solution creep; (b) An apparatus that enables the evaporating solvent to condense back, and the supply of nutrient that keeps the solution concentration steady; the rotation allows uniform growth of the crystal initiated on the seed crystal. Drawn schematically based on Figure 9 and 20 in ref. 6.

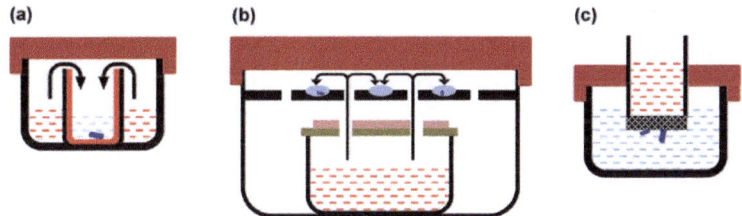

Figure 7.13 Crystal growth by non-solvent addition/diffusion; the solution (light blue), non-solvent (red), and crystals forming (dark blue plates) are indicated in each case: (a) a simple apparatus where the non-solvent vapors condense into the solution kept in an inner beaker; the silanized wall (red lining) inhibits solution creep; (b) an apparatus that enables the non-solvent vapor that diffuses through controllable vents to condense into sitting drops of the solution; (c) an arrangement that allows the non-solvent to diffuse through a membrane into the solution. The crystals formed in each case are represented by dark blue plates.

Addition or diffusion of a non-solvent (for the material of interest) into a nearly saturated solution of the material is another commonly employed method; the non-solvent should be miscible with the solvent used to prepare the solution. Addition of the non-solvent reduces the solubility of the solute at the temperature of the experiment, and causes precipitation; if carried out in a controlled manner at a slow rate, it can lead to the formation of good quality crystals.

- The simplest apparatus involves a beaker of solution kept surrounded by an easily vaporizable non-solvent within a closed environment; vapors of the non-solvent gradually condensing into the solution causes the crystal growth (Figure 7.13(a)).
- A more controlled environment and evaporation/condensation/crystallization can be achieved using the so-called 'sitting drop' method (Figure 7.13(b)).
- Diffusion of the non-solvent into the solution across a membrane filter is yet another variation that can be used to grow crystals in the solution (Figure 7.13(c)). The diffusion can also be into a gel medium (in which case the membrane may not be required); the gel environment can facilitate uniform crystal growth.

7.3.3 Cooling Methods

Simple cooling reduces the solubility of the solute in most cases, and causes crystal growth. In many cases it is seen that, if the solubility is in the range 5–30 wt%, and the solubility decreases with temperature, a slow cooling of 1–5° can produce a success rate of 90% for the crystallization.

Cooling in a gel medium is often employed because of several advantages it offers. A commonly used gel is produced by the hydrolysis of tetraalkoxysilanes (~5–10 vol%). Nucleation occurs on supercooling, but can be retarded by increasing the temperature. Advantages of the crystal growth in a gel environment include:

- the avoidance of convection currents in the solution, and growth by diffusion alone,
- uniform growth of the crystal suspended in the gel, with all faces receiving the nutrients (different from that of a crystal growing at the bottom of a beaker),
- formation of fewer aggregates and effective isolation of each growing crystal.

A **laminar flow system** in which the solution flows over the seed crystal without turbulence or uncontrolled convection currents, provides a sophisticated approach to crystal growth by cooling (Figure 7.14). The propulsion helix circulates the solution, the diaphragm and the conical section facilitate the laminar flow over the seed, rotated to facilitate uniform growth. The temperature control allows gradual cooling. In some cases, the flow rate can be used to control the crystal growth rate, achieving fast crystal growth in selected materials. In materials which show highly anisotropic growth,

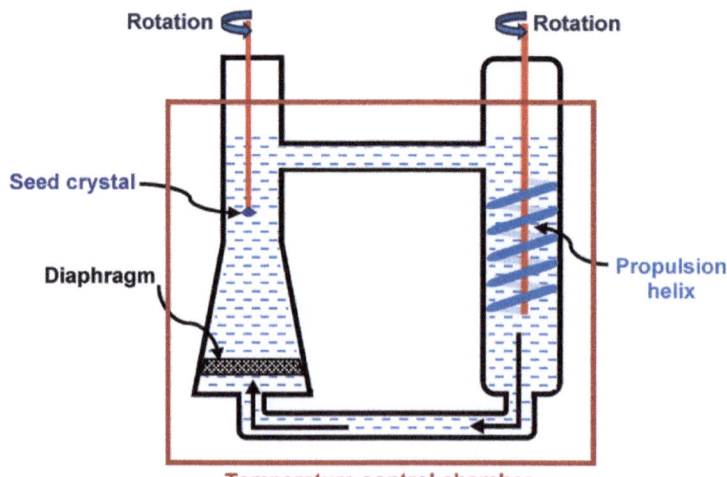

Figure 7.14   Crystal growth by cooling in a laminar flow system.

directed flow can enhance the deposition on slowly growing crystal faces.

7.3.4 Hydrothermal Process

When the material of interest has very low solubility in available solvents, and tends to decompose if high temperatures are needed either to dissolve or melt, hydrothermal methods come in useful.

- High temperature and pressure conditions in closed environments, typically in an autoclave, are employed.
- The material of interest and water are enclosed in a sealed container (typically with Teflon lining that provides a non-reactive enclosure). This is placed in a protective mesh or chamber and heated to the required temperature; the water turns to vapor and the pressure within the system increases accordingly.
- Dissolution occurs under these hydrothermal conditions, and subsequent cooling leads to crystal growth.

Examples of materials that are crystallized using the hydrothermal method include sapphire (Al_2O_3), garnet ($Y_3Fe_5O_{12}$), and β-quartz. β-quartz is an illustrative case. It decomposes if heated above 575 °C; so melt-growth is not advisable. It does not dissolve in any solvent up to 200 °C or so. However, it dissolves in water at its critical temperature of 374 °C, and high pressure, ~218 atm. Crystallization is achieved after the dissolution under such special conditions. Alkaline solutions under the hydrothermal conditions can also be employed.

The hydrothermal process is often used to carry out chemical reactions, and crystals of the product formed under those conditions. Use of non-aqueous media is also common; such processes are described as **solvothermal.**

7.3.5 Derivatization Cases

In several instances, products of a chemical reaction, derivatization, complexation, *etc.*, are directly formed as crystals, by controlling the rate at which the reaction proceeds and the product assembles and grows.

- A typical example is the formation of salts of acids and bases. Combining an amine and a mineral acid under controlled conditions leads to crystals of the salt resulting from the protonated amine and the conjugate base of the acid. This is facilitated by the proper choice of solvent in which the salt is relatively less soluble than the amine and the acid. In some instances, the mineral acid can be taken in water, and the amine in an organic medium that is immiscible with water; when the two solutions are in contact, crystals of the salt can form at the interface.

- Crystals of donor–acceptor complexes are often prepared by allowing the donor and acceptor molecules to diffuse and meet. If the solvent is chosen such that the two components are soluble, but not the complex, the slow diffusion and combination can produce crystals of the complex. A simple apparatus that facilitates this process is shown in Figure 7.15(a); the donor (D) and acceptor (A) are taken in two flasks which are connected by an inverted Y-shaped tube. The common solvent is added, the system closed by a stopper at the top, and maintained at constant temperature. Slow dissolution and diffusion of D and A molecules leads to their complexation at the junction and crystallization of D–A.

Electrocrystallization is ideally suited when an oxidation/reduction process leads to the formation of an insoluble product, the crystals of which are of interest. Electrochemical oxidation/reduction leads to the formation of a cation/anion close to the anode/cathode, and combination with the counterion available

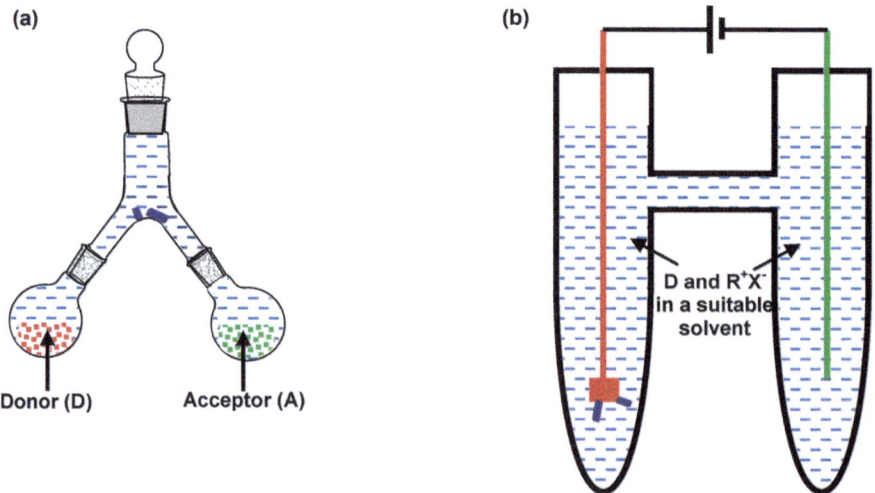

Figure 7.15   (a) Growth of D–A crystals by diffusion of D and A in solution. (b) Electrocrystallization of D^+X^- on the anode following oxidation of D in presence of the electrolyte R^+X^-. The crystals formed in each case are represented by dark blue plates.

(often from the electrolyte) leads to the product crystal growing on the electrode.

- The cases of organic conductors and semiconductors are well known; see the examples of organic superconductors given in Table 4.5.
- In a typical experiment, neutral molecules of the donor (D) are taken with an electrolyte R^+X^-, mostly in a non-aqueous solvent. On application of the appropriate potential for the electro-oxidation of D, D^+ ions are formed at the anode; they form a complex with X^- from the electrolyte, and D^+X^- crystals grow on the electrode (Figure 7.15(b)).
- As D^+X^- is often very insoluble, normal modes of recrystallization are impossible.
- The electrically conducting nature of D^+X^- ensures that the crystal continues to grow as an extension off the electrode.

7.4 Crystal Growth: Gas, Liquid, Solid States

As discussed in the previous section, crystal growth from solution is the most common in chemistry, especially when dealing with molecule based materials. On the other hand, crystals of solids such as metal oxides, ceramics, and various inorganic semiconductors, are grown from melts, directly from vapors and occasionally through solid state transformations.

7.4.1 Sublimation Methods

The simplest mode of crystal growth by sublimation involves heating a solid (crude product with impurities), and condensing the vapors on a cooled surface (cold finger) to obtain crystals (of the pure material). A reduced pressure facilitates sublimation of the solid instead of melting to form the liquid and subsequent evaporation (recall single component phase diagrams). A simple apparatus used is shown in Figure 7.16(a).

A **Knudsen cell** allows combination of two reactants to form crystals of the product (Figure 7.16(b)). The components are heated in separate cells (maintained at different temperatures if required); the vapors meet and condense to form crystals of the complex on the cold finger.

Methods like chemical or physical vapor deposition (CVD, PVD), molecular beam epitaxy, *etc.*, are techniques used to prepare crystalline thin films. These will be discussed briefly in Section 7.5.

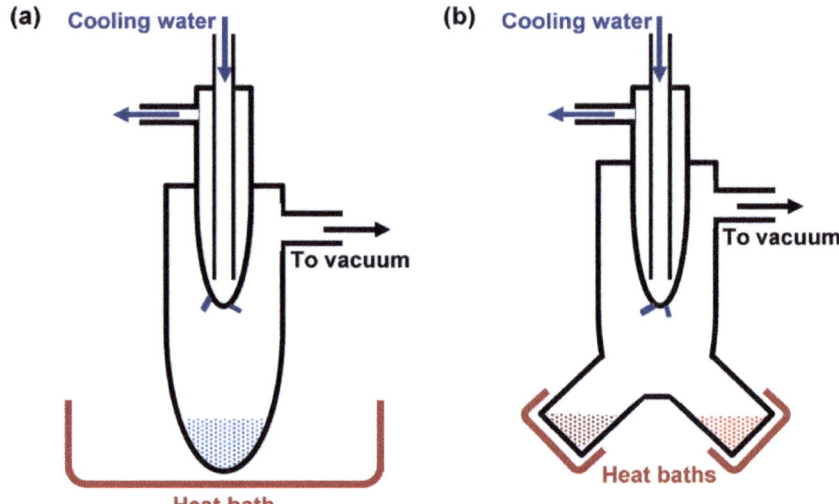

Figure 7.16 Schematic diagram of the (a) setup for crystal growth by simple sublimation of a solid, and (b) Knudsen cell for sublimation of two solids to form crystals of their adduct or complex. The crystals formed in each case are represented by dark blue plates.

7.4.2 Melt Growth Techniques

Crystals of several technologically important materials, such as semiconductors, are grown from their melts, silicon being one of the most famous examples. Controlled cooling of the melt can lead to crystallization. For melt growth techniques to be useful the material should be stable against decomposition upon melting.

The melting point of the material of interest can be brought down by adding foreign substances. The added substance is technically called a flux; it is essentially a molten solvent from which the material of interest crystallizes out upon cooling. An example is the case of potassium titanyl phosphate ($KTiOPO_4$) grown from a polyphosphate ($K_6P_4O_{13}$) flux.

The **Bridgman–Stockbarger** method involves the slow movement of the melt through a temperature gradient across the melting point of the material of interest (Figure 7.17(a)).

- The temperature in the two zones (T_1, T_2) are controlled separately, and maintained above and below the melting point (T_M), i.e., $T_1 > T_M > T_2$.
- The conical tip with small volume helps to reduce the number of nuclei formed, as the melt moves into the zone with temperature below the melting point.

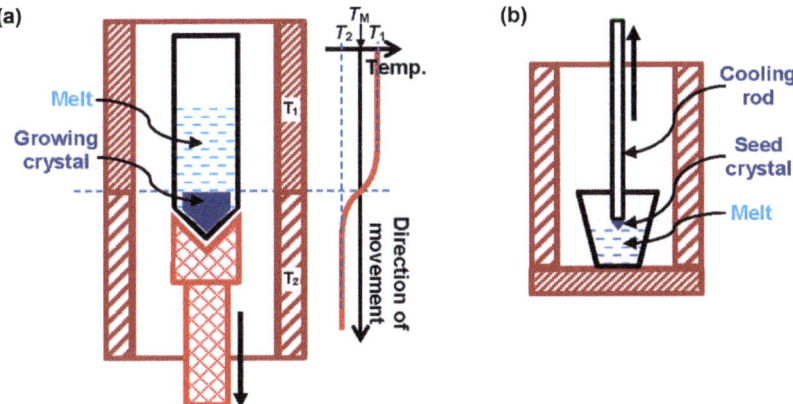

Figure 7.17 Schematic representation of the techniques for crystal growth from the melt: (a) Bridgman–Stockbarger (the zone temperatures, T_1 and T_2 maintained above and below the melting temperature of the material, T_M, respectively), and (b) Czochralski (crystal pulled up, out from the melt at the end of a cooling rod).

- A seed crystal at the tip can enable crystal growth with a preferred orientation.

The **Czochralski (crystal pulling) method** uses a seed crystal attached at the end of a cooling rod (maintained at a temperature below the melting point), pulled slowly out from the melt; the side and bottom of the chamber can be maintained at different temperatures (Figure 7.17(b)).

- The rate of pulling is decided based on factors such as the thermal conductivity and latent heat of fusion of the material, and the rate of cooling of the rod.
- Special morphologies of crystals can be realized using this method.

7.4.3 Crystallization in the Solid State

Devitrification is a process in which crystals grow from the glassy or supercooled liquid state of a material.

- With many complex molecules, the temperature at which the nucleation occurs efficiently can be well below the temperature at which the crystal growth is fastest.
- The material in the supercooled state is heated close to or just above the melting temperature, cooled slowly, and the cycle repeated, until the crystal growth (often through heterogeneous nucleation) occurs.

Crystals of many ceramic materials are grown through the **sol–gel** route. A colloidal solution (the sol) under suitable conditions forms a network of discrete particles or polymers (the gel). The technique is described in Section 7.6.2 in the context of nanomaterials. Emergence of crystals from the gel network involves significantly milder conditions of temperature, pressure, *etc.*, compared to the conventional ceramic routes.

Other processes that are of interest for crystal growth starting from the solid state include zone-refining and sintering, discussed in the following sections.

7.4.4 Zone-refining

Zone-refining is a method used to obtain highly pure crystalline materials, commonly used for purifying metals and semiconductors. The method exploits the difference in the solubility of the impurities in the melt state and the solid form. The basic idea of zone-refining is the following:

- Assume that the impurity is more soluble in the melt than in the solid, *i.e.,* the equilibrium constant, $K = \dfrac{c_{\text{melt}}}{c_{\text{solid}}} \gg 1$, where c_{melt} and c_{solid} are the concentrations of the impurity in the melt and solid respectively.
- Melt a small zone of a solid ingot or rod of the impure material, and then run this zone along the length of the sample.
- The impurity distributes preferentially into the melt, is transported and accumulates at the end leaving the earlier parts in a pure state.
- Multiple runs of the zone movement across the length of the ingot enable iterative purification, leading to a highly pure final product.

The process is illustrated schematically in Figure 7.18. By cutting off the regions with the impurities accumulated, a pure crystalline rod of the material is obtained.

7.4.5 Sintering

Sintering is the process in which powder particles bind together when maintained at a temperature slightly below the melting temperature of the material. It is often used to prepare larger crystalline particles from smaller ones, and is technologically important due to the impact on various materials properties like mechanical hardness, thermal

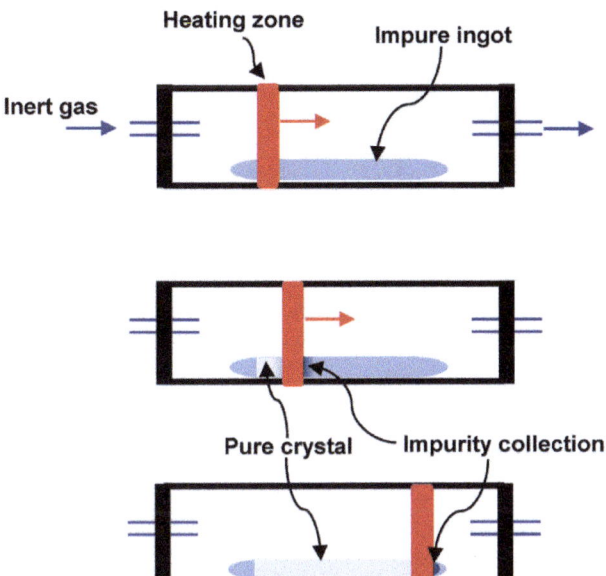

Figure 7.18 Stages (top to bottom) in zone-refining. A heating coil passes from left to right over the impure ingot placed in an inert gas atmosphere; the heating zone moves along leaving pure crystalline material behind and accumulating the impurities in the molten region.

and electrical conductivity, ductility, transparency, coercive force, magnetic permeability, *etc.*

- The driving force for the sintering to occur is the decrease of surface free energy (arising from unsatisfied surface bonds).
- The stages in the sintering process that occur due to atomic or ionic motion are schematically illustrated in Figure 7.19(a). The particles start joining through 'necks' and the domains start fusing together; the last stage shown in the figure has two domains with some pore structure. The ideal final state would be a continuous defect-free crystal.
- The mass transport in the solid state occurs through:
  - Diffusion: slow across the volume, faster at grain boundaries and fastest at the surfaces; a model for the diffusion process is shown in Figure 7.19(b).
  - Flow: which could be described as viscous in the amorphous phases and plastic in the crystalline regions; dislocations cause the flow.
  - Evaporation–condensation process: especially in materials with a high vapor pressure and close to the melting temperature.

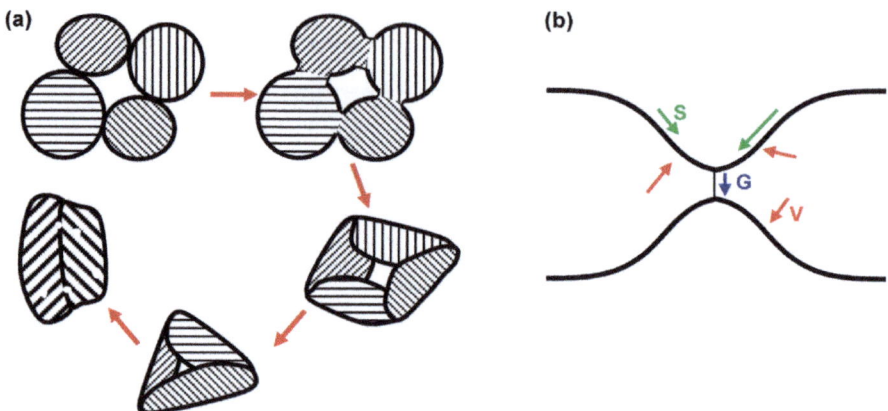

Figure 7.19   (a) Schematic diagram showing the stages in a sintering process. (b) A model for the diffusion paths (V=volume, G=grain boundary, S=surface) involved in the initial stages of sintering. Drawn schematically based on Figures 17 and 18 in ref. 1.

The rate of sintering is affected by factors like the particle size distribution, surrounding atmosphere, non-stoichiometry, presence of impurities and defects, electrical field (in the case of ferroelectric materials), magnetic field (in the case of magnetic materials), *etc.*

- Defects generally enhance the sintering rate.
- Impurities can influence the sintering rate in different ways. Aliovalent impurities like Li^+ in ZnO create O^{2-} vacancies and increase the sintering rate, but Al^{3+} in ZnO decrease the O^{2-} vacancies and the sintering rate.
- The rate of sintering is inversely related to the particle size, *i.e.*, the rate $\propto \dfrac{1}{r^3}$ where r is the average particle radius. An example is the sintering of alumina (Al_2O_3): particles obtained by grinding have sizes of the order of 3 μm whereas those obtained from calcination have sizes typically 0.03 μm. The rate of sintering of the latter particles is ~10^6 times faster than that of the former.

7.5   Thin Films

Thin films of materials, typically a few micrometers in thickness, are extremely important in various technologies and applications. Typical examples of their applications are:

- electronic circuits
- semiconductor devices

- display systems
- sensors and switches
- Josephson devices
- as protective layers (*e.g.*, high speed steel cutting tools) and for decorative purposes.

Various kinds of substrates are used for depositing and growing thin films. These include glass, polycrystalline ceramics, selected single crystal surfaces, and metals.

Several techniques have been developed for the fabrication of thin films. The simplest procedures are based on evaporation of solutions, commonly used in the case of polymers; spin-coating is a particularly versatile approach. Vapor deposition methods (physical and chemical) are often employed; these processes may involve thermal evaporation or sputtering. Other approaches utilize chemical or electrochemical reactions.

During or after thin film deposition, its thickness is monitored using a variety of methods.

- Quartz crystal microbalance (QCM) is a commonly used device. Thin film deposition on the QCM changes its mass and hence resonance frequency. The mass deposited is estimated from the change in the resonance frequency; knowing the area of deposition and density of the material, the thickness of the thin film can be determined.
- The change in optical transmittance can be used to determine the thickness, based on Beer–Lambert's law.
- Other measurements that are used include electrical resistance, capacitance, *etc.*

7.5.1 Spin-coating

Evaporation of a drop of solution on a substrate leaves behind the solute in a solid form. In the case of solutes such as polymers, thin films are commonly formed. This process is called **drop-casting**.

If the substrate is dipped in the solution, and the solution film evaporated subsequently, the process is known as **dip-coating**. A continuous and uniformly thin film can be formed by spinning the substrate when the drop of solution placed on it is evaporating. This is known as **spin-coating**.

- The standard spin-coating setup is shown schematically in Figure 7.20; the substrate is held on the spinning platform by

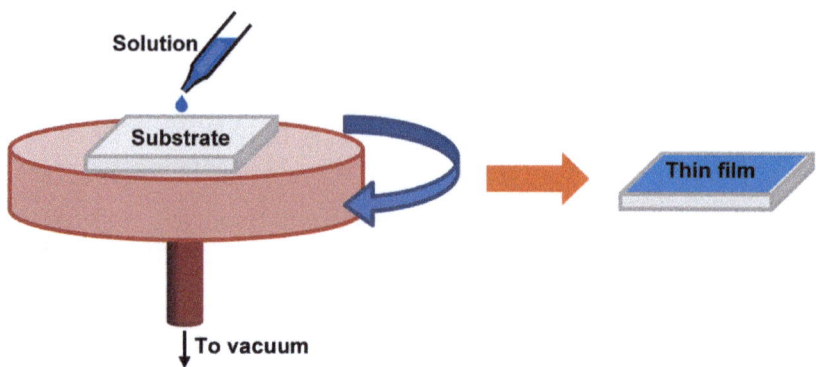

Figure 7.20   Schematic representation of the spin-coating process; the substrate is usually held on the spinning platform by vacuum suction through a hole in the center of the platform connected to the pump through the spinning axis.

vacuum suction through a hole in it (vacuum chuck). The solution spreads uniformly by centrifugal force on the drop during spinning.

- The important parameters that influence the spin-coating process and the quality of the thin film obtained are the concentration and viscosity of the solution, vapor pressure of the solvent, and the spinning conditions.
- Higher spinning speeds and lower solution viscosity lead to thinner films.
- Factors such as temperature and atmospheric condition under which the process is carried out can be fine-tuned to modify the spin-coated film properties.

7.5.2   Thermal Evaporation

Thermal evaporation is the most popular method of physical vapor deposition (PVD) of thin films. A common example is the formation of metal thin films, for example, as electrodes for electrical measurements on solid samples.

- The metal source is taken either as a filament or in a suitable non-reactive container (usually a Mo, W, or Ta boat) which can be heated electrically. When the temperature reaches the evaporation point, the metal atoms leave the source and condense on a suitable substrate maintained at a lower temperature, to form the thin film (Figure 7.21(a)).
- Aluminum is a typical example; heating at ~1220 °C provides a vapor pressure of ~0.01 Torr, leading to aluminum thin film deposition.

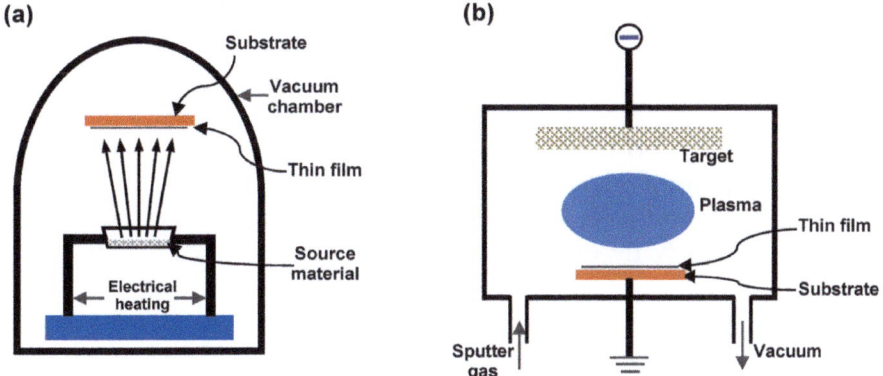

Figure 7.21 Schematic diagram of the setup for (a) thermal and (b) sputter deposition of thin films.

Reactive evaporation involves chemical reaction along with evaporation; for example, heating a metal in the presence of oxygen, leads to the formation of the metal oxide thin film. A typical example is: $2Al + 3O_2 \rightarrow Al_2O_3$.

Two-source evaporation is commonly employed to prepare semiconductor thin films, for example, InAs or CdSe. In specialized flash evaporation processes, a single heating filament with the required elements can provide thin films, *e.g.,* Ni–Fe–Cr film.

Electron beams or laser beams can be used to evaporate the source materials and deposit thin films. The latter is called **laser ablation**; a typical example is the fabrication of thin films of the superconductor $YBa_2Cu_3O_{7-x}$, using a Nd:YAG laser on the bulk material in an O_2/Ar atmosphere with controlled partial pressure of O_2. **Pulsed laser deposition (PLD)** uses a high power pulsed laser on a target material, causing electronic excitation followed by the formation of a plasma plume, and evaporation or ablation of the material.

7.5.3 Sputtering

In the method described in the previous section, heat is used to eject atoms from the target material. In sputtering, energetic ions are used instead; it is a plasma based deposition technique.

- The target (source of the material of interest) and substrate (on which the thin film is formed) are enclosed in a vacuum chamber and a voltage applied so that they form the cathode and anode respectively (Figure 7.21(b)).

- A sputter gas (typically Ar) is sent into the chamber at an appropriate pressure.
- Stray high energy electrons from the cathode accelerate towards the anode, collide with the gas atoms, ionizing them $(e^- + Ar \rightarrow Ar^+ + 2e^-)$. Formation of additional electrons leads to a cascading process, and plasma production.
- The energetic ions in the plasma striking the target liberate its atoms which fly out and deposit on the substrate forming the thin film.

In **magnetron sputtering**, magnetic fields are used to trap electrons near the target and increase the efficiency of the initial ionization. Magnetically confined plasma forms at lower gas pressures. Charge build-up on the target can be avoided by **RF (radio frequency) sputtering**, alternating the bias voltage on the target and substrate at radio (MHz) frequencies.

When separate ion sources are employed, the process is known as **ion beam sputtering**. Some sputtering processes involve chemical reaction of the target atoms with the ions leading to thin films of the reaction product. Typical examples are of metals with ions of oxygen or nitrogen to form oxides and nitrides: Ar/O_2 on Al leading to thin films of Al_2O_3, Ar/N_2 on Ti leading to TiN thin films, *etc.* These are referred to as **reactive sputtering** processes.

7.5.4 Chemical and Electrochemical Methods

Chemical vapor deposition (CVD) involves the decomposition and/or reaction of one or more volatile precursors followed by deposition on a substrate to form the thin film. The reaction could be of different types.

- Oxidation, reduction, nitriding, *etc.*: *e.g.*, (i) $nM + (m/2)G_2(g) \rightarrow M_nG_m$; (ii) Si_3N_4 thin films grown using silane (SiH_4) or hexamethyldisiloxane as the source of Si and NH_3 as the source of N.
- Disproportionation: *e.g.*, Ge reacted with I_2 at 400 °C forms GeI_2; it disproportionates $(2GeI_2 \rightarrow GeI_4 + Ge)$ to deposit Ge thin films on a substrate at 210 °C.
- Polymerization: monomers in the gas phase produced by UV or electron beam irradiation on the source material, deposit and polymerize to form polymer thin films on the substrate.

Metal organic chemical vapor deposition (MOCVD) uses metal alkyls or metal alkoxyls as precursors, which after vaporization, get pyrolyzed and deposit as metal thin films on the substrate. Thin films of compound semiconductors are often prepared using this method.

- Examples of precursors are, various metal acetylacetonates [$M(acac)_n$], titanium tetraisopropoxide [$Ti(OPr^i)_4$], *etc.*
- A well-known example is the reaction of trialkyl gallium and arsine in presence of H_2 gas to form GaAs thin films: $(CH_3)_3Ga + AsH_3 \xrightarrow{H_2} GaAs + 3CH_4$.
- Deposition on a well-defined semiconductor surface, and growth of the film in registry with the substrate structure is called epitaxy. In the MOCVD deposition, it is called **metal organic vapor phase epitaxy (MOVPE)**.

The **sol–gel process** is often employed to grow thin films. Typically, metal alkoxides are hydrolyzed, coated on a surface, gelled, and dried. An example is a silica (SiO_2) coating formed from tetraethoxysilane, $(C_2H_5O)_4Si$.

Several electrochemical techniques are used to fabricate thin films.

- **Electroplating** is employed to deposit thin films of metals like Ag or Cu, by electrochemical reduction of the cations from the corresponding electrolytes, at the cathode. This process can be carried out also by normal chemical reduction using suitable reducing agents, *e.g.,* formaldehyde to reduce $AgNO_3$ to form Ag film.
- **Anodization** is used to grow thin films of metal oxides on metal (M) surfaces (*e.g.,* Al_2O_3 on Al). The general reactions that occur at the anode and cathode are $M + nH_2O \rightarrow MO_n + 2nH^+ + 2ne^-$, and $2ne^- + 2nH_2O \rightarrow 2nOH^- + nH_2 \uparrow$ respectively.

7.5.5 Patterning of Thin Films

Thin films with specific patterns are required in devices fabricated for various applications in optics, electronics, opto-electronics, and sensing. Many specialized techniques are employed to create patterned thin films; some of the general and popular approaches are listed here. The method chosen depends on the nature of the materials as well as the size and type of patterns that need to be created.

Lithography is a commonly used method; photolithography and electron beam lithography are often employed to create patterns on the micro- or nanoscale.

Photolithography involves the use of light (of suitable wavelength, often UV) irradiation to transfer a geometrical pattern design from a

mask to a photosensitive material (**photo-resist**), which in turn allows patterning of the thin film on the substrate.

- The mask could be made of glass, polycrystalline ceramics, single crystals, metals, *etc.*, on which the required geometric pattern is formed, typically using electron beam or laser irradiation.
- The photoresist is usually an organic compound. Negative resists become less soluble where irradiated; commonly, a monomer gets polymerized. Positive resists become more soluble where irradiated; typically, photochemically generated acids cleave bonds.
- There are primarily three steps involved in photolithography: coat, expose, and develop (Figure 7.22).
  - First, a resist is coated on top of the thin film (on a suitable substrate) in which the pattern is to be formed.
  - Covering with the patterned mask, it is exposed to light of suitable wavelength; due to a photochemical process, the exposed material becomes more (positive resist) or less (negative resist) soluble than the unexposed material.
  - In the 'develop' stage, the soluble resist material is washed away using a suitable solvent, and the exposed regions of the thin film etched (*e.g.*, using HF for metal oxides). On stripping the remnant photoresist material, the pattern becomes visible. The pattern from the mask is now transferred on the substrate as the corresponding pattern of thin film (with negative resists) or the corresponding gap in the thin film (with positive resists).

Electron beam lithography is used when high resolution patterns are required; the significantly smaller wavelength of the electron beam compared to light, allows patterns with sizes less than 10 nm to be fabricated.

7.6 Colloids and Nanomaterials

Colloids are homogeneous dispersions of small (a few nm to several μm) particles of a substance in the continuous medium of another substance; this is distinct from a solution in which there is complete mixing of the two at the atomic/molecular level.

- Depending on the phases involved, the colloid can be a sol (solid in liquid), emulsion (liquid in liquid), foam (gas in liquid), gel (liquid in solid), *etc.*

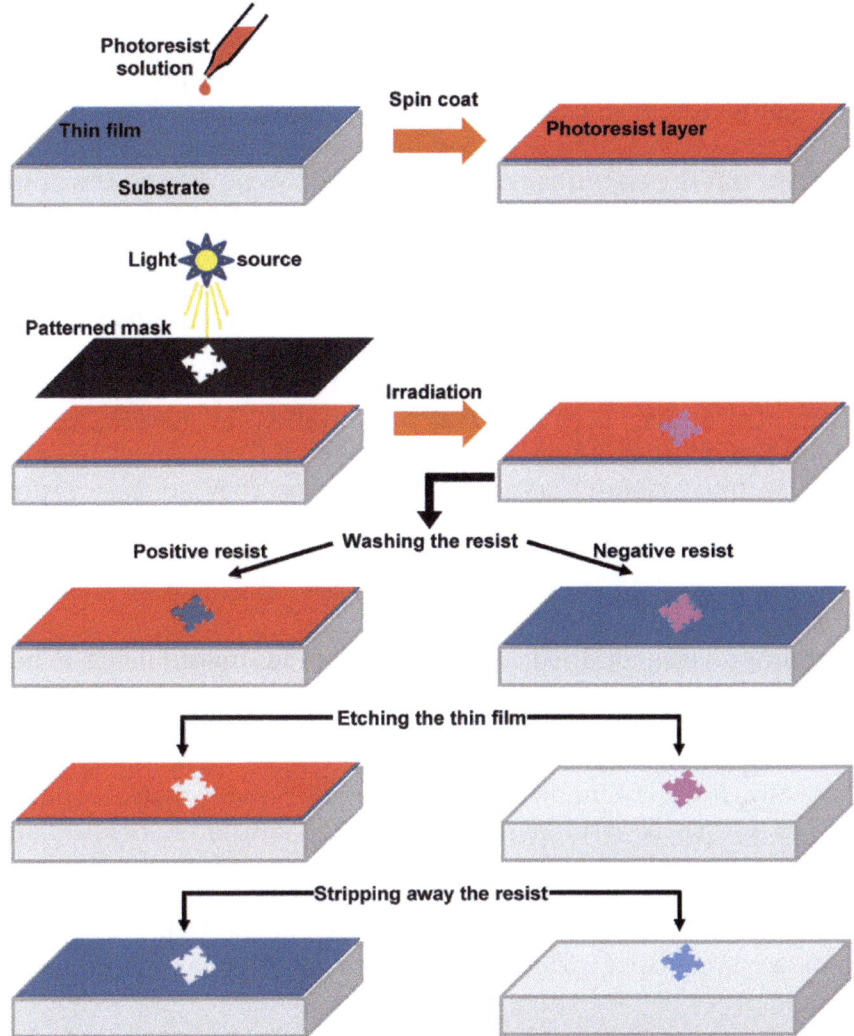

Figure 7.22 Stages in photolithography (top to bottom). *Coat*: spin coating the photoresist on the thin film (on substrate). *Expose*: photo-irradiation through the mask with a pattern (irradiated region of photoresist chemically changed). *Develop*: washing of the resist (positive/negative: exposed region more/less soluble); exposed thin film etched; remaining photoresist stripped away. Pattern formed in the thin film (right, with negative resist), or gap in the thin film (left, with positive resist).

- In typical colloids (milk being a common example), the particle sizes are large enough to scatter visible light, making it translucent or even opaque.

- If the particle sizes are significantly smaller than the wavelength range of visible light, the scattering will be negligible, and the dispersion transparent; such a system is sometimes described as a 'nanoparticle solution'.
- Terms like 'a colloidal solution of nanoparticles' may be found often; but the significance of size should be clear from the points noted above.

Nanoparticles have sizes typically in the range ~1–100 nm along all three dimensions (Figure 7.23); in the case of semiconductors these are called **quantum dots**. If the nanometric size restriction is along only two dimensions, with the length in the third dimension being significantly larger, they are called **nanorods** or **nanowires**. 2-D structures with the length in one dimension alone in the nano regime, would naturally be called **nanosheets**; they are also termed **ultrathin films** (fabrication discussed in Section 7.7).

- **Nanocomposites** are bulk materials formed with nanostructures embedded within.
- Another class defined as **nanostructured materials**, are bulk materials having an internal structure with characteristic length scales of the order of 1–100 nm.

Critically, the term 'nano' becomes relevant in a materials context, when the properties and/or functions of the substance in the nano-metric size regime:

- are different from that of the substance at the bulk level, and
- show a systematic variation with size within the nanometric range.

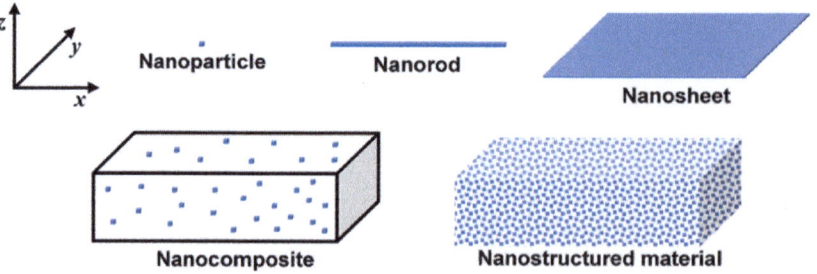

Figure 7.23   Schematic diagram of a nanoparticle (length along x, y, z~1–100 nm), nanorod (length along y, z~1–100 nm), nanosheet (length along z~1–100 nm), nanocomposite and nanostructured material.

These unique aspects make nanomaterials of great interest and utility in many applications. The important point is that beyond chemical composition and structure, **size becomes a parameter** that can be tuned to vary materials properties. Many of the unique features arise due to the large surface area to volume ratio.

- In the case of spherical particles the ratio is inversely related to the radius; a consequence is the enhanced localized surface plasmon resonance (LSPR) extinction (Section 6.1.2).
- The large surface area facilitates enhanced surface activities such as adsorption and catalysis.

The **confinement** of electrons and other charge carriers to small spaces leads to prominent size-dependent optical and electrical properties (quantum confinement phenomena). Small sizes can have a significant impact on the magnetic domain structure in magnetic materials. Mechanical properties of materials are also often different in the nano regime, as the defect distributions and crystalline phase boundaries are altered.

Nanoclusters are particles with sizes typically <2 nm; in some important aspects, they show properties distinctly different from that of the relatively larger nanoparticles. For example, noble metal nanoclusters can exhibit strong fluorescence emission, but no LSPR; the corresponding nanoparticles can show LSPR, but no fluorescence emission.

There are innumerable methods for the synthesis of nanomaterials and fabrication of nanostructures.

- In a general sense, materials with nanometric (10^{-9} m) size can be synthesized or fabricated by (i) assembling atoms and molecules, typically with dimensions in the Å (10^{-10} m) range, or (ii) breaking down bulk materials, typically in the mm–cm (10^{-3}–10^{-2} m) or higher size range. These are popularly called the 'bottom-up' and 'top-down', or 'build-up' and 'break-down' approaches respectively (Figure 7.24).
- The broad principles of these general methods of nanomaterials fabrication are noted in the following sections; nanoparticles, nanowires, and nanocomposites are covered here. Monolayers and ultrathin films are discussed in Section 7.7.

7.6.1 Top-down Approaches

In order to dissolve solids in a liquid, a common practice is to first grind the solid into a fine powder; the smaller particle size and increased

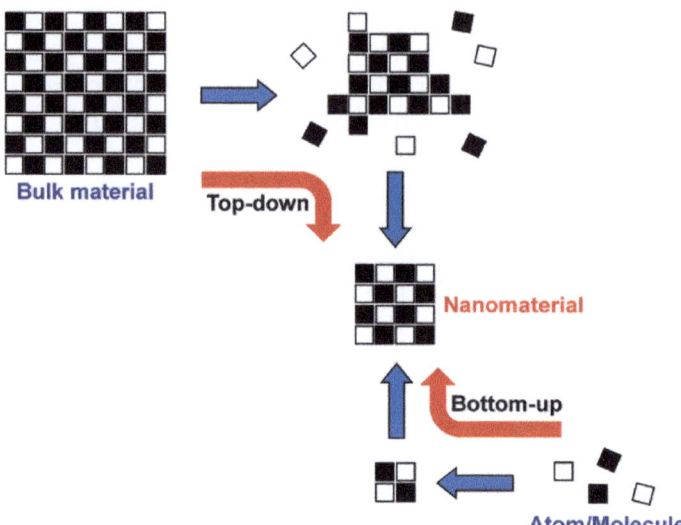

Figure 7.24 Schematic representation of the 'top-down' and 'bottom-up' fabrication of nanomaterials (not to scale).

surface area enable easier dissolution. A similar principle applies to the various 'top-down' approaches to nanomaterials synthesis.

Ball-milling is a well-known traditional method to break down solids into a fine powder or to blend solid particles.

- It is a mechanical process in which the material is ground by collision with hard balls or pebbles (often made of stainless steel or ceramic materials) when the two are confined in an enclosure and rotated at high speeds.
- The critical speed of the grinding mill corresponds to the rotation that generates centrifugal forces equaling gravitation force so that the balls do not fall off the shell of the mill; the ball mills are operated at 65–75% of the critical speed.
- High energy ball-milling involving high energy collisions between the solid material and the balls, is used to make nanomaterials; a classic example is the production of fine and uniform dispersions of Al_2O_3, Y_2O_3, ThO_2, *etc.*, in Ni based superalloys (capable of efficient operations at high temperatures).
- High energy ball-milling can also lead to chemical reactions between solids being ground together; this is called mechanochemistry.

Lithography in a general sense, is the technique in which a pattern prepared on a flat surface (stone [*lithos*] originally, with oil for patterning) attracts the ink allowing the faithful printing of the pattern on a substrate like paper.

- Short wavelength light and electron beams enable patterning at micro- and nanoscales.
- Photolithography and electron beam lithography are popular top-down approaches to create nanostructures. The light or electron beam can remove materials or write patterns on polymer coatings by inducing chemical reactions in the irradiated region. Section 7.5.5 provides a detailed outline of the patterning procedure in the context of thin films.

Exfoliation is the separation of bulk material having a layered structure, into mono- or few-layer structures. The latter, having only a few atoms bonded in the direction perpendicular to the layer, would obviously be nanometric in thickness; they are called nanosheets. The famous examples in this area include graphene and MXenes, discussed in Section 7.7.4.

Laser ablation synthesis in solution involves laser irradiation of a substrate (commonly a metal plate) kept within a liquid; the plasma plume formed by the atoms removed from the substrate (ablation) condenses to form nanoparticles.

7.6.2 Bottom-up Approaches

The bottom-up approaches exploit weak (various non-covalent) or strong (covalent, ionic, *etc.*) interactions between the building blocks of atoms or molecules to assemble them into nanostructures.

- The basic processes are similar to those involved in crystallization (Section 7.2 to 7.4). The critical issue is to stop the growth of the particle at nanometric scales, so that bulk material formation is avoided. This can be achieved by controlling concentrations, growth conditions, or the introduction of external agents (often called the capping agent) that prevent continued growth.
- Depending on whether the growth is arrested in all three directions, only two, or just one, the resulting nanostructure would respectively be a nanoparticle/quantum dot, nanowire/nanorod, or nanosheet/nanoplate.

The **colloidal route** is the most facile and popular method for the bottom-up synthesis of nanomaterials. Many kinds of nanomaterials can be synthesized through this approach; a few well-known examples are listed below.

- Noble metal nanoparticles can be prepared by the 'polyol method'. It involves the chemical reduction of the precursor metal ions by alcohols, which in turn get oxidized to aldehyde or ketone. Ethylene glycol is a typical reducing agent; precursors for Ag and Au nanoparticles are usually $AgNO_3$ and $HAuCl_4$. If a suitable capping agent (*e.g.*, poly(vinylpyrrolidone)) is present, the metal atoms formed grow into nanocrystals; bulk metal is not formed. A fine illustration of this involves the Tollens reagent. Treatment of a glucose solution with ammoniacal $AgNO_3$ leads to the formation of a silver mirror (bulk Ag); however, if this is carried out in the presence of starch (as the capping agent), a yellow solution is obtained (containing Ag nanoparticles).
- The classic example of colloidal synthesis is the experiment by Faraday; treatment of $HAuCl_4$ with phosphorus in carbon disulfide, produced what he called 'finely divided gold', which are now described as nanoparticles.
- Many metal oxide nanoparticles are prepared by the hydrolysis of the metal salt in an alkaline medium, followed by thermal decomposition of the hydroxide.
- Control of nucleation and growth by parameters like concentration, pH, temperature, ultrasonication, or microwave irradiation, allows modification of the size and size-distribution, shape, and even the crystal structure (in some cases) of the resulting nanoparticles.

Reverse-micelles with a hydrophilic interior and hydrophobic shell provide a confined aqueous space for the formation and growth of nanoparticles. Control of the micelle size and shape allows tuning of the nanoparticle characteristics. Surfactant molecules such as cetyl-trimethylammonium bromide (CTAB, cationic) and sodium bis(2-ethylhexyl)sulfosuccinate (Aerosol-OT or AOT, anionic) are commonly used. Many metal oxide nanoparticles have been synthesized through these routes. The surfactants help to stabilize micro-emulsions (mixtures of immiscible liquids), enabling the mixing of the reagents introduced into the two immiscible media and precipitation of the nanoparticle product. The protocol is sometimes referred to as the **micro-emulsion** technique.

Sol–gel synthesis provides a route for the preparation of various oxide nanoparticles. Silica and titania are well-known examples; hydrolysis of tetralkoxy silane (*e.g.*, $Si(OEt)_4$) or titanium tetraalkoxide (*e.g.*, $Ti(i\text{-}PrO)_4$) followed by polymerization, produces the sol and gel respectively. The gel is dried and subjected to thermal treatments like

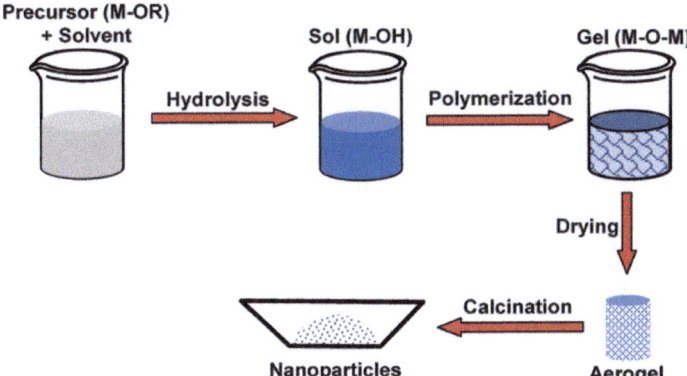

Figure 7.25 Steps involved in the sol-gel synthesis of oxide (MO) nanoparticles. The precursor alkoxide (M–OR) on hydrolysis produces M–OH which on polymerization forms M–O–M; extension of M–O–M leads to the gel network of MO.

calcination, to generate the nanoparticles; the general scheme is shown in Figure 7.25.

Methods like **CVD**, discussed in Section 7.5.4 in the context of thin films, can be used to fabricate nanoparticles; the deposition time and experimental conditions can be tuned in order to limit the size of particles being deposited to the nano regime. Many semiconductor nanoparticles are fabricated by the CVD and related methods. **Electrodeposition** methods can also be employed in a similar manner.

DNA origami (origami ≡ Japanese art of paper folding) is a specialized technique used to build nucleic acid-based nanostructures. The self-recognition property of DNA and its assembly into duplex and other secondary structures are exploited to fabricate unusual nanostructures that can function as biological scaffolds, nanofilters, *etc.*

7.6.3 Some Specialized Nanomaterials/Nanostructures

Core–shell nanoparticles have an inner core and outer shell made up of distinct materials or components.

- Core–shell structures serve special functions. A typical utility is in catalysis, with the more active (often expensive) shell material and the less active (cheaper) core material, providing a large active surface area at a relatively lower total cost. The core can also influence the electronic structure and hence catalytic activity of the shell material. Often the shell serves as a protective layer for a reactive or toxic core.

- They can be fabricated in sequential steps. A typical example is $Au@SiO_2$ (Au core inside silica shell) particle, prepared by citrate reduction of $HAuCl_4$ to form Au nanoparticle, followed by sol–gel synthesis of the silica shell.
- CdSe@ZnS particles can be prepared by first forming CdSe particles by the micro-emulsion method using $Cd(ClO_4)_2$ and $(Si(CH_3)_3)_2Se$, followed by precipitating ZnS shell also by the micro-emulsion method using $Zn(ClO_4)_2$ and Na_2S.
- Many particles have a polymer shell. $TiO_2@PS$ can be synthesized by a sol–gel process for the core followed by ligand exchange for the polystyrene shell.

Nanotubes and nanowires are of particular interest in various devices, especially due to the electronic and excitonic transport in 1-D.

- Spontaneous assembly and highly anisotropic 1-D growth can lead to nanowires or nanotubes. Carbon nanotube (CNT) is a famous case; among the many methods for growing CNTs, catalytic chemical vapor deposition is the most common. Hydrocarbons (*e.g.*, acetylene) are passed over Fe or Co catalyst at high temperatures (600–1200 °C); the molecule decomposes and CNT grows off the catalyst surface.
- In the **vapor–liquid–solid (VLS)** method, vapors of the material of interest collect on catalyst droplets on a substrate, and grows at the liquid–substrate interface, forming the nanowires (the catalyst may remain at the tip) (Figure 7.26(a)). Nanowires of several semiconductors like GaAs, ZnS, *etc.*, and binary alloys like SiGe are grown using the VLS method.
- Various template-based methods can be used to grow nanowire structures. The templates can be polymers like DNA or nanopores in a membrane.
- Anodization (electrochemical oxidation, Section 7.5.4) of aluminum can produce nanoporous alumina membrane (anodic aluminum oxide, AAO). The pores are then filled with the material of interest; metals can be deposited by methods like chemical vapor deposition or electrochemically, whereas polymers are deposited by filling with the solution and drying. When the membrane template is dissolved in alkali, the nanowire array is obtained; the substrate can also be etched or polished away, to release free nanowires. The overall process is shown schematically in Figure 7.26(b).
- **Electrospinning** is an elegant method to produce nanowires or nanofibers in large quantities. A solution (mostly of polymers) is

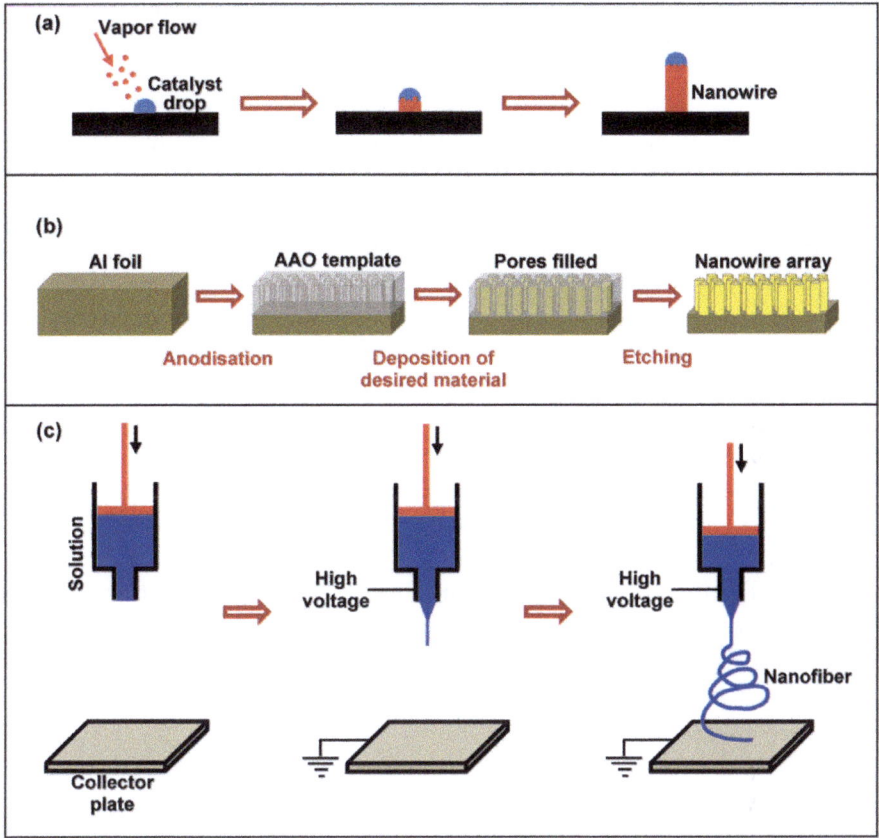

Figure 7.26 Schematic representation of the (a) VLS method to form nanowire, (b) formation of an AAO template and nanowire array fabrication, and (c) electrospinning method in which nanofiber is deposited on a collector plate.

pumped through a spinneret (spinning tip) with a high voltage applied on the tip (Figure 7.26(c)). The solution gets charged; when the repulsive electrostatic force exceeds the surface tension, the droplet is stretched into what is known as the Taylor cone (middle stage in Figure 7.26(c)). As the stretched droplet extends, the solvent evaporates and the solid fiber is deposited on the grounded collector plate; diameter of the fibers are typically a few hundreds of nanometers. Small bends in the fiber lead to spiraling structures.

Nanocomposites contain more than one chemical component, with at least one of them in the nanometric size range. There are innumerable combinations which give rise to nanocomposites; the typical example of metal nanoparticles in a polymer matrix can be used to illustrate some general fabrication routes.

- An obvious method is to prepare the metal nanoparticles (using the methods discussed earlier), and then mix them in solution state with the polymer solution to form the composite. The mixture can then be cast into a thin film by spin-coating. Homogeneity of the nanocomposite film will depend on the extent of mixing achieved.
- Plasma deposition involves the formation of a plasma from the metal target, together with the polymer or its precursor also converted into the vapor form. The two co-deposit on a substrate in the form of a nanocomposite thin film. The process involving plasma generation, may require relatively high power.
- *In situ* synthesis is a facile method to prepare polymer–metal nanocomposite thin films. A typical procedure to prepare supported or free-standing Ag–PVA (silver nanoparticles in poly(vinyl alcohol)) thin film is shown in Figure 7.27. First, a thin film of polystyrene (PS) is spin-coated on the substrate (*e.g.,* a glass plate). A mixture of aqueous solutions of PVA and AgNO$_3$ in the required proportion is then spin-coated on the PS layer. Upon heating (typically below 100 °C for less than 1 h), the Ag$^+$ ions get reduced by the hydroxyl groups on PVA to Ag atoms, which assemble to form Ag nanoparticles within the polymer film. The polymer serves as the reducing agent for the formation of nanoparticles, as well as the stabilizer for the nanoparticles formed. The nanocomposite film with the PS support can be peeled off the substrate; if dipped

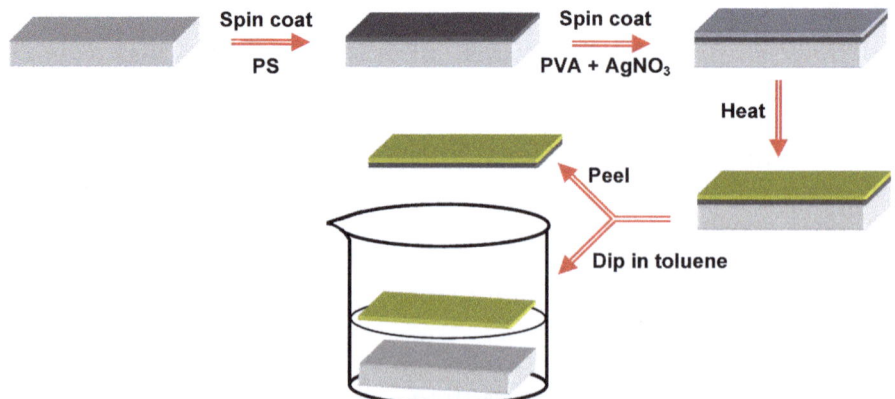

Figure 7.27   Schematic diagram of the *in situ* fabrication of polymer–metal (*e.g.,* Ag–PVA) nanocomposite thin film.

in toluene, the PS layer dissolves, releasing the Ag–PVA as a free-standing film, typically a few tens of nanometers thick.

7.7 Monolayers and Ultrathin Films

Ultrathin films, typically a few or tens of nanometers in thickness, are made up of a single layer or a few layers of atoms or molecules. The extremely small thickness can lead to attributes that are unique to nanostructures. The large effective surface area and porosity enable highly efficient sensing responses in these materials.

Like the other nanomaterials discussed in the previous section, these can also be fabricated in a bottom-up fashion or top-down mode. The former can be through self or steered assembly of the building blocks. Some selected methodologies are discussed in the following sections.

7.7.1 Self-assembled Monolayer

Ordered monolayer structures formed on clean surfaces by the adsorption of surfactant molecules are called self-assembled mono-layers (SAM). The classic example is the assembly of alkanethiols on a crystalline gold surface. When a gold surface, typically (111), is exposed to di-n-alkyl disulfides in solution, the latter undergoes S–S bond cleavage and forms strong Au–S bonds; this generates an ordered SAM of the corresponding n-alkylsulfide on the surface. Similar SAMs can also be fabricated by exposing the surface to alkane thiols in solution. With S as the anchoring group, the alkyl chains may be chosen with appropriate functionality as the terminal group at the other end, in order to fabricate SAMs with surface functionality. Formation of such SAMs is shown schematically in Figure 7.28. Depending on the choice of the organic molecule and surface, temperature, *etc.*, the reaction times required for SAM formation could vary from a few minutes to several hours.

Another well-known example of SAM formation is by the adsorption of chlorosilanes or alkysilanes on glass. They form covalent bonds with the silica surface leading to very stable SAMs. This is a common method to make the glass surface hydrophobic.

Fatty acids can form bonds with metal atoms on a surface to produce SAMs. A typical example is that of docosanoic acid [$CH_3(CH_2)_{20}COOH$] on silver oxide; ionic bonding between the carboxylate group and the metal cation provides stability.

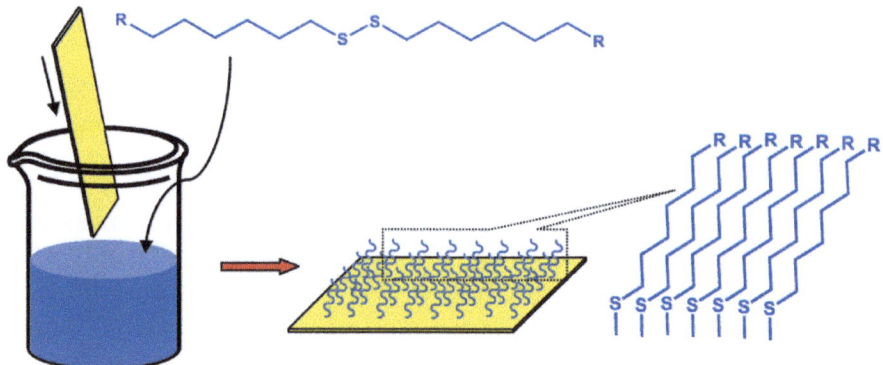

Figure 7.28 Schematic diagram showing the formation of a SAM. A gold substrate (yellow) is dipped into a solution (blue) of the di-*n*-R-alkyldisulfide (with terminal R groups), to produce a well-ordered monolayer assembly of R-akylsulfide on the substrate.

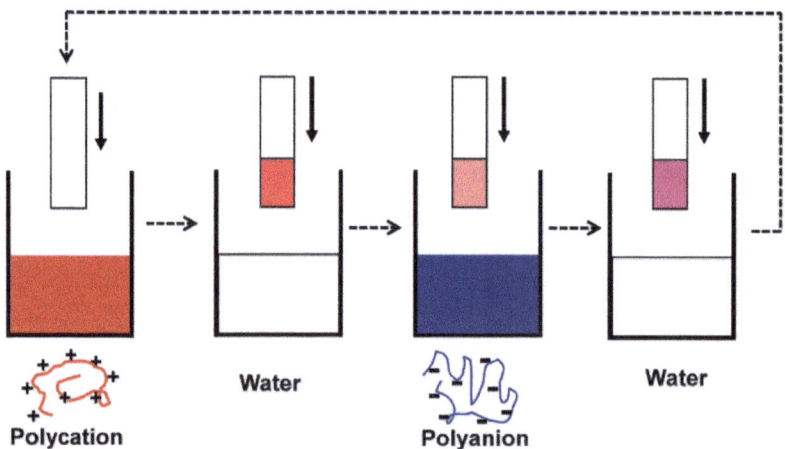

Figure 7.29 Schematic representation of the steps in the layer-by-layer assembly of a polycation and polyanion, with water wash in between.

7.7.2 Layer-by-layer Assembly

The technique involves alternate deposition of oppositely charged molecules, polymers, nanoparticles, *etc.*, on a suitable substrate with wash steps in between, to build multilayer films in a highly controlled fashion. The deposition could be by simple methods such as immersion of the substrate in the appropriate solutions, spin or spray coating. The in-between washing is to remove material deposited beyond a single layer.

A well-known example of layer-by-layer (LbL) assembly involves alternate deposition of a polycation and a polyanion (Figure 7.29).

Inclusion of layers of metal nanoparticles with surface charges is easily realized in the LbL assembly. Linear growth of the thickness and related properties (for example optical absorption) can be achieved by systematic increase of the number of layers.

7.7.3 Langmuir–Blodgett Film

The Langmuir–Bodgett (LB) technique developed by Irvin Langmuir and Katherine Blodgett is one of the most elegant methods to fabricate ultrathin films of surfactant molecules, polymers, functionalized nanoparticles, *etc.*, with control on the packing density and assembly. It is a fine example of mechanically steered assembly that leads to mono- or multilayer ultrathin films with varying thickness.

The main steps involved in LB film fabrication are listed below, followed by more detailed discussion of each step.

- Spreading of the surfactant molecule on the surface of pure water (other liquids are rarely used) taken in an LB trough (Figure 7.30).
- Mechanical compression of the monolayer of surfactant molecules at the air–water interface, to a desired close-packed 2-dimensional assembly.
- Transfer of the monolayer on to a suitable substrate by vertically dipping it in and out through the monolayer (transfer by contacting the monolayer horizontally with the substrate is also known, and is called the Langmuir–Schaefer method).

One of the most popular surfactants used in LB film fabrication is an amphiphilic molecule; an amphiphile has a hydrophilic head group and hydrophobic tail structure (example: stearic acid

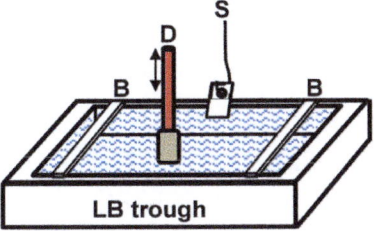

Figure 7.30 Schematic picture of an LB trough (usually made of an inert and hydrophobic material like Teflon) with high purity water in it. **B**s are movable barriers which can slide over the water surface. **S** is a surface pressure sensor (often using a filter paper in contact with the water surface, known as a Wilhelmy plate, hung from a sensing unit). **D** is a dipper that can move a substrate in and out through the water surface.

(Figure 7.31(a))). The amphiphilic molecules are spread on the surface of highly pure water taken in the LB trough, by dropping a known volume of a solution (with well-defined concentration) of the molecule in a volatile organic solvent (Figure 7.31(b)). The solvent evaporates and the molecules, being surfactants, spread on the aqueous surface; the hydrophilic head is solvated by water whereas the hydrophobic tail sticks out.

- The surface tension of water (γ_0) in contact with the air, at ambient temperature (25 °C) is 72.8 mN m^{-1}. It decreases to γ due to the surfactant molecules.
- The decrease is known as the surface pressure, defined as, $\pi = \gamma_0 - \gamma$.

The monolayer is compressed using the mechanical barriers, **B** in Figure 7.30 (Figure 7.31(c), (d)). The process at the surface can be

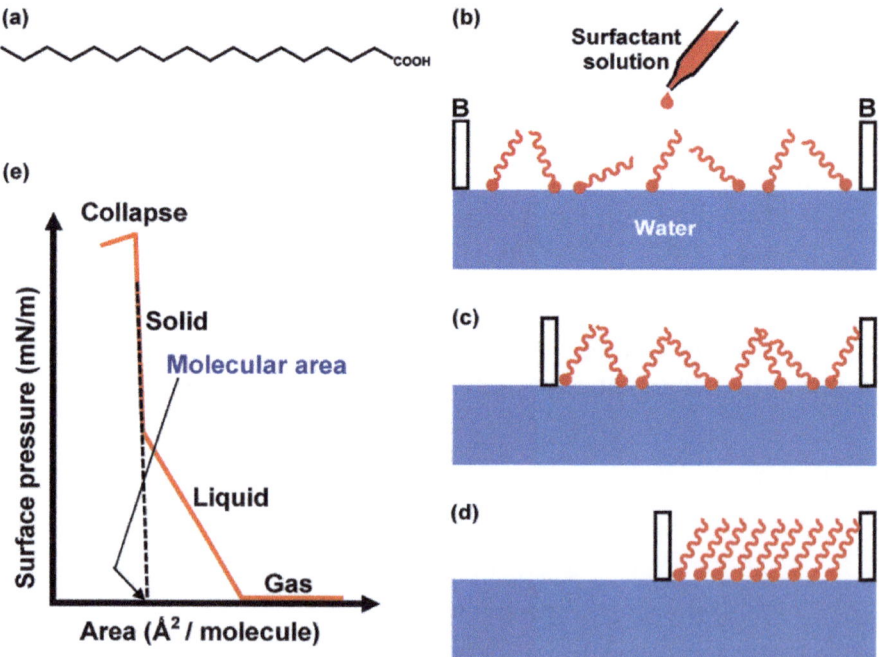

Figure 7.31   (a) Molecular structure of stearic acid. (b) Spreading the surfactant solution on water surface in an LB trough; the 'gas' state of the monolayer formed at the air–water interface. (c, d) Compression of the monolayer to the 'liquid' and 'solid' states. (e) An idealized surface pressure–area isotherm showing the different monolayer phases, collapse point, and extrapolation to the molecular area.

monitored by a plot of the surface pressure (π) *versus* area per molecule (A) (Figure 7.31(e)).

- A is estimated using the area over which the molecules are spread and the exact number of molecules (from the concentration and volume of the solution spread).
- The π–A isotherm (at a selected temperature) is the 2-D analog of the pressure–volume plots (at constant temperature) for gases.
- Three regions can generally be identified in the π–A isotherm (Figure 7.31(e)). At very low or zero surface pressure, the molecules are far apart; this is defined as the 'gas' state. On compression, the molecules start contacting and the pressure increases with a low slope, in the 'liquid' state. When the molecules are close packed, they form an ordered 2-D lattice; the pressure increases sharply due to the reduced compressibility. The state is defined as the 'solid'.
- Extrapolation of the 'solid' domain of the isotherm to zero surface pressure provides a direct estimate of the molecular area as shown in Figure 7.31(e).
- On continued compression, the ordered monolayer can break or in some cases, reorganize into a multilayer structures; this is defined as the 'collapse' region.
- Other kinds of structural phase changes in the monolayer (for example, reorientation of the head group) can show up in the π–A isotherm, in the form of slope changes, formation of plateaus, *etc.*

When a substrate with a suitable surface is dipped through the compressed monolayer at the air–water interface (usually in the 'solid' state), a single layer can get transferred on to it. The process of first and second layer deposition is shown in Figure 7.32(a), (b).

- While dipping a substrate with a hydrophobic surface into the monolayer, the hydrophobic tail groups get attached. If the surface is hydrophilic, when the substrate is drawn out, the hydrophilic head groups get attached.
- Repeated dipping leads to multilayer deposition (Figure 7.32(c)). With specialized amphiphiles and substrates different types of multilayer structures can be made.

Common examples of amphiphiles fabricated as LB films include fatty acids such as stearic and arachidic acids, corresponding alcohols, various molecules of interest such as dyes functionalized with

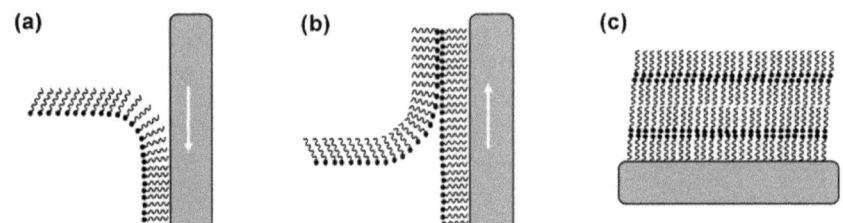

Figure 7.32   (a, b) Schematic representation of the deposition of a monolayer at the air–water interface on to a substrate (hydrophobic surface) dipped vertically in and out through the interface. (c) A 4-layer film formed on the substrate. Note that the monolayer and film are significantly thinner than the substrate; the figure is obviously not to scale.

long hydrocarbon chains, amphiphilic polymers, nanoparticles capped with molecules possessing long alky chains, *etc.*

LB films are unique materials having an organized assembly that can be steered through careful control of the monolayers at the air–water interface, and systematic tuning of the thickness and relative orientations of multilayer stacks. They find extensive applications as model systems to understand various interfacial phenomena, and device applications in molecular electronics, non-linear optics, sensing, and device development.

7.7.4   2-D Materials: Graphene and MXenes

2-D materials are of interest in various applications like electronics, sensing, and catalysis. A famous example is **graphene**. Graphite with the weak binding between the sheets made up of hexagonal carbon rings is easily cleaved, forming graphene, which consists of a single or few layers of carbon. It has unique electronic properties due to the delocalized π-electronic structure of the honeycomb lattice.

Other well-studied nanosheets include MoS_2, WS_2, $MoSe_2$, graphitic carbon nitride (g-C_3N_4), hexagonal boron nitride (h-BN), *etc.* **MXenes** is a general name used to denote 2-D (nanosheet) materials with the general formula, $M_{n+1}X_nT_x$ (M = Ti, V, Cr, Mo, W, *etc.*; X = C or N; T = O, F, Cl, functional groups like OH, *etc.*; $n = 1, 2, 3$).

Some common methods for the fabrication of 2-D materials are the following.

- Chemical vapor deposition (Section 7.5.4) is often employed for fabricating graphene. The precursor is pyrolyzed to form C atoms, which are then assembled on a substrate under high temperature. Catalysts are useful in both stages; in the second

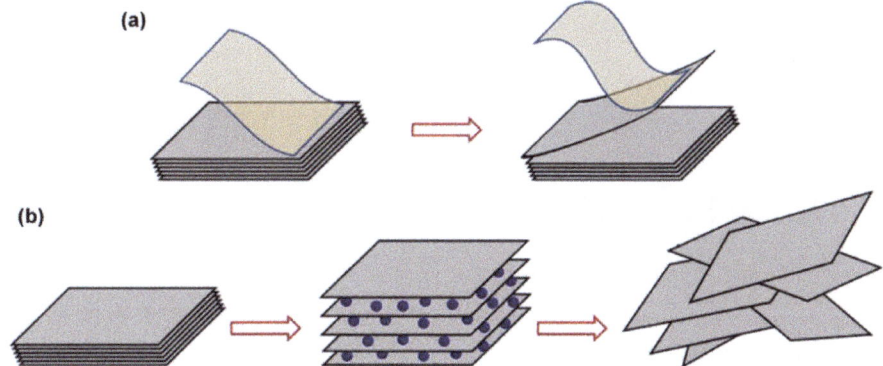

Figure 7.33 Exfoliation of a layered material like graphite by (a) mechanical peeling using a Scotch tape, and (b) intercalation of guest molecules or particles (blue spheres) followed by thermal or chemical treatment, to form nanosheets.

stage, copper substrate used as the substrate can serve as the catalyst.

- Exfoliation is a unique method for forming 2-D materials; it can be carried out in many ways, mechanically, thermally, by chemical means using intercalation, *etc.* The simplest mechanical approach involves sticking Scotch tape on a piece of the layered material and peeling off the layers successively (Figure 7.33(a)).
- Intercalation of guest atoms or molecules into the interlayer space can be used to weaken the layer-to-layer interaction and separate them into 2-D nanosheets (Figure 7.33(b)); the latter step can be achieved thermally or chemically.

References

1. H. V. Keer, Solid State Transformations, Reactions and Crystal Growth, in *Principles of the Solid State*, Wiley Eastern Limited, 1993, ch. 7.
2. J. L. MacManus-Driscoll, *Adv. Mater.*, 1997, **9**, 457.
3. V. Ramamurthy and K. Venkatesan, *Chem. Rev.*, 1987, **87**, 433.
4. G. Wegner, *Die Makromol. Chem.*, 1972, **154**, 35.
5. P. G. Vekilov, *Nanoscale*, 2010, **2**, 2346.
6. J. Hulliger, *Angew. Chem., Int. Ed.*, 1994, **33**, 143.

8 Exercises

8.1 Chapter 1

A. Content Based

1. Explain in detail why rotation symmetry operations of order 1, 2, 3, 4, and 6 only are compatible with a periodic lattice.
2. List the Bravais lattices arising from the cubic, tetragonal, and orthorhombic systems. Discuss their genesis and account for why the cubic system has three, the tetragonal system has two, and the orthorhombic system has four Bravais lattices.
3. Cubic close packed structure and simple hexagonal lattice are both Bravais lattices, but a hexagonal close packed structure is not. Explain using a consideration of primitive vectors.
4. Write down a set of primitive vectors for the following Bravais lattices: (i) face-centred cubic (fcc), (ii) body-centred tetragonal (bct), and (iii) edge-centred orthorhombic.
5. Write down the vector for the translation along the three edges of an fcc unit cell, in terms of the primitive vectors defined in Exercise 4.
6. Derive the reciprocal lattice vectors corresponding to the primitive direct lattice vectors defined in Exercise 4.
7. Show how the von Laue condition for X-ray diffraction relates to the Bragg equation.
8. Using schematic diagrams of the NiAs and CdI_2 structures, describe how the $1:1$ and $1:2$ stoichiometries are satisfied, and explain the similarity between the two.

Core Concepts for a Course on Materials Chemistry
By T. P. Radhakrishnan
© T. P. Radhakrishnan 2023
Published by the Royal Society of Chemistry, www.rsc.org

B. Context Based

1. Determine the efficiency of packing of spheres in primitive cube, fcc and bcc lattices (packing efficiency is the ratio of the volume occupied by the spheres in a unit cell to the total volume of the unit cell).

2. In a system of cubic close packed (fcc) spheres, determine the ratio of the radius of the tetrahedral interstitial site to the radius of the octahedral interstitial site.

3. (i) A 2-D rectangular lattice with unit cell lengths, a and b is shown in Figure 8.1; draw a set of primitive vectors, a_1 and a_2 on this lattice, and write them in terms of the Cartesian axes. (ii) Indicate the unit cell lengths of the reciprocal lattice shown in Figure 8.1; draw the vectors, b_1 and b_2 on the lattice, and write them in terms of the Cartesian axes.

4. Draw a schematic diagram of the 2-D hexagonal lattice; write down a set of primitive lattice vectors a_1 and a_2 in terms of the unit cell length a, and indicate them on the diagram. Indicate the reciprocal lattice vectors b_1 and b_2 on the diagram.

5. Given the equation for the geometric structure factor, $S_{hkl} = \sum_n f_n e^{i\pi(hx_n+ky_n+lz_n)}$, derive the systematic absence conditions for a face-centered cubic lattice.

6. Write down the direct and reciprocal lattice vectors for diamond considering it as an fcc lattice with two atoms in the basis. Write down also the coordinates of the two atoms in the basis. Determine the systematic absence conditions for its X-ray diffraction profile.

7. Consider the centered rectangular lattice (a 2-D Bravais lattice) shown in Figure 8.2. (i) Write down a set of primitive lattice vectors and the corresponding reciprocal lattice vectors. (ii) Derive the structure factor, S_{hk} for this lattice and determine the systematic absence condition.

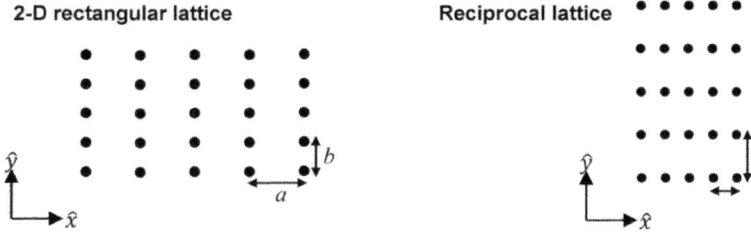

Figure 8.1 Schematic drawing of the 2-D rectangular lattice and the reciprocal lattice (Exercise 8.1.B3).

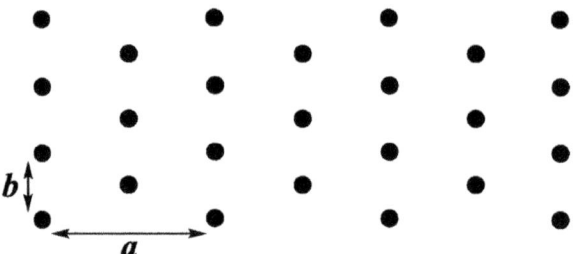

Figure 8.2 Schematic drawing of the centred rectangular lattice (Exercise 8.1.B7).

8. Describe the spinel and inverse spinel structures. Give examples of materials possessing such structures.
9. Draw schematic diagrams to illustrate the following structures: rock salt, cesium chloride, fluorite, and rutile.
10. What is the difference between the structures of α-graphite and β-graphite?
11. Contrast the zinc blende and wurtzite structures of ZnS; state the similarities and differences.
12. Diamond has a zinc blende structure – explain.
13. Illustrate with a diagram, the perovskite structure for the general oxide formula ABO_3. What is the oxygen coordination for A and B? Indicate this on the diagram.
14. Describe how liquid crystals represent a phase in between liquids and crystals, and state the relevance of order in terms of orientations and dimensionality with a brief outline of different classes of liquid crystals.

8.2 Chapter 2

A. Content Based
1. State whether the densities of crystals are affected by the presence of Schottky and Frenkel defects. If they are, how?
2. The addition of Ca^{2+} increases the conductivity of a KCl crystal. However, it is found that the conductivity is due to the motion of K^+ ions rather than the dopant Ca^{2+} ions. Discuss the mechanism for the enhanced conductivity, with a suitable diagram.
3. If the ratio of A^{3+} to A^{2+} in a non-stoichiometric sample of an ionic crystal of the oxide, AO is 0.15, what is the nature of the defect? Derive the correct formula for AO.
4. In the non-stoichiometric metal oxide, $M_{1+x}O$, which kind of ion, in which type of site, forms the dominant defect?

5. In a sample of $YBa_2Cu_3O_{7-\delta}$ it is found that 10% of the copper is in the $+3$ oxidation state and the remainder is in the $+2$ state. Estimate the value of δ.

6. Discuss the nature of non-stoichiometry in (i) copper(I) sulfide, (ii) zinc oxide, and (iii) praseodymium oxide. What are the different structures in the case of (iii)?

B. Context Based

1. The average energy needed to create a vacancy in a monatomic solid is 1.15 eV mol^{-1}. Compute the ratio of the number of vacancies at 1000 K and 500 K.

2. An element with atomic weight 40.0 g mol^{-1} forms a bcc lattice with unit cell length 4.0 Å. If it has 0.2% Schottky defects, what is its density in kg m^{-3}?

3. If 5% of the K^+ ions in a KCl crystal are replaced by Ca^{2+} ions, what is the change in the density of the crystal? [Atomic weight $(g\,mol^{-1})$: $K = 39.10$; $Ca = 40.08$; $Cl = 35.45$.]

4. Suggest the possible formula for a charge compensation defect material produced by incorporating Eu^{3+} ions in ZnO, simultaneously creating a vacancy in the cation sublattice. What properties are likely to be of interest in these materials?

5. Tungsten oxides can exist as non-stoichiometric compounds, WO_{3-x}. Write the formula showing the distribution of oxidation states of W.

6. Analysis of a sample of titanium oxide showed the formula to be $TiO_{0.8}$. Explain the origin of this non-stoichiometric composition and estimate the detailed formula showing the composition of the ions in different oxidation states.

7. A tungsten bronze crystal has the formula $Na_{0.1}WO_3$. In what states is tungsten present and in what ratio?

8. Some specific structures of zirconia (ZrO_2) can be stabilized by adding small percentages of yttria (Y_2O_3) or calcia (CaO). What are the general compositions of the yttria-stabilized zirconia (YSZ) and calcia-stabilized zirconia (CSZ)? What kinds of defects are created in these mixed metal oxides? What is the percentage of these defects, if 10% of Y_2O_3 or CaO are added? What physical property of the material is likely to be influenced by these defects?

9. Draw a schematic diagram to illustrate the vacancies in some of the Cu–O planes of $YBa_2Cu_3O_{7-\delta}$ for $\delta = 0.5$, 1.0, in comparison with the case when $\delta = 0$.

8.3   Chapter 3

A. Content Based

1. What are the numbers of acoustic and optical branches in the phonon dispersion plots of a 1-D crystal lattice with three atoms in the basis?
2. If the length of a 1-D crystal is L, what are the possible phonon wavelengths?
3. Write down the Bose–Einstein distribution function which gives the probability of excitation of phonons with frequency ω at temperature T, and show that it is proportional to T at temperatures that are high compared to the phonon energies.
4. Define Einstein and Debye temperatures of solids. State the significance of the latter by comparing the Debye temperatures of any two solids.
5. List the main assumptions used in the derivation of Debye's model for heat capacity of solids. What is the temperature dependence of heat capacity well above the Debye temperature and at low temperatures?
6. Describe the role of the anharmonicity of lattice vibrations in the thermal expansion and thermal conductivity of solids.

B. Context Based

1. Draw a schematic diagram to show the optical and acoustic phonon modes with $\lambda = 4a$, in a 1-D lattice with unit cell dimension a, and a 2-atom basis.
2. Draw a schematic diagram of the phonon dispersion curve for a 1-D crystal lattice with a monatomic basis of H atoms; label the axes and relevant points on the axes. Indicate on the same diagram, the phonon dispersion curve for a similar lattice of D atoms (assume that the force constants are the same for the H and D lattices).
3. Using a schematic diagram of a 1-D crystal lattice with unit cell of length a, and a monatomic basis, illustrate a longitudinal phonon with $\lambda = 8a$. On a plot of the phonon dispersion curve, indicate the wave vector and frequency corresponding to this phonon mode.
4. Derive a simple expression for the total energy of a longitudinal wave in a monatomic 1-D lattice of atoms with mass M and nearest neighbor interaction force constant, C.
5. What would be the effect on the heat capacity of metals at room temperature if, in an imaginary sense, the Planck's constant

were to be increased and Boltzmann's constant reduced ten-fold each?

6. Why is the heat capacity of diamond considerably less than 3R at room temperature?

7. Derive the expression for the density of states for phonons in a 2-D lattice.

8. Derive the full expression within the Debye model, for the heat capacity valid at any temperature.

9. If the Debye frequency of solid A is twice that of solid B, calculate the ratio of the heat capacities C_A/C_B at any temperature T (in the low temperature regime).

10. Similar to the expression for the linear thermal expansion coefficient, α_L, write down the expression for the area thermal expansion coefficient, α_A. Show that, for a small extent of isotropic expansion, α_A is twice α_L, and hence also proportional to the heat capacity.

8.4 Chapter 4

A. Content Based

1. Using Fermi–Dirac statistics, show that $\mu = \varepsilon_F$ at 0 K; explain the physical meaning of this equation.

2. Heat capacity of free electron gas is $\sim 1\%$ of that expected on the basis of the law of equipartition of energy. Why?

3. What are the basic assumptions of the free electron theory? What are the phenomena explained by it? And what are its main failures?

4. Write the equation for σ (electrical conductivity) in terms of n (the number density of electrons), e and m (charge and mass of the electron respectively), and τ (relaxation time for electron scattering); using appropriate units of n, e, m, and τ show that σ has the unit $\Omega^{-1} \mathrm{m}^{-1}$.

5. Thermal conductivity of a metal is 406 $\mathrm{W\,m}^{-1}\mathrm{K}^{-1}$ at 300 K. Using Wiedemann–Franz law, estimate its electrical conductivity.

6. Write expressions to show how the heat capacity of insulators and metals vary with temperature at low temperatures. What is the reason for the difference between the two?

7. Given the Bloch function, $\psi_k\left(\vec{r}\right) = u\left(\vec{r}\right).e^{i\vec{k}\cdot\vec{r}}$, obtain the eigenvalue of the crystal translation operation $\vec{T}$ (a lattice vector translation).

8. The band gap of Si and Ge are 1.12 eV and 0.66 eV respectively. Draw schematic plots of the variation of their electrical conductivity (σ) with temperature (T), highlighting the comparison between the two.

9. Which of the oxides, FeO, Fe_2O_3, Fe_3O_4, and $NiFe_2O_4$ is expected to show the highest hopping electronic conductivity? Explain why.

10. Explain qualitatively, the origin of isotope effect in superconductors.

B. Context Based

1. Draw a schematic diagram to show the Hall effect (in the standard geometry, thin rectangular plate) with holes as the charge carrier.

2. Resistivity of a metal is 10^{-6} Ω m. If it has 10^{28} electrons m^{-3}, calculate the relaxation time for electron scattering within the free electron model.

3. For a simple cubic lattice, show that the kinetic energy of the free electron at a corner of the first Brillouin zone (i.e. having wave vector at this point) is higher than that of an electron at the midpoint of a face of the zone. What bearing does this have on the conductivity of divalent metals?

4. Derive an expression for the electronic density of states $\mathfrak{D}(\varepsilon)$ for a 2-D free electron gas.

5. Draw qualitative band diagrams for sodium, magnesium, and diamond indicating the valence atomic orbitals involved in the band formation, to provide a simple view of their electrical properties.

6. Experimentally, MnO is found to be a semiconductor. Draw a qualitative band diagram for MnO; is it expected to be a semiconductor on the basis of this band picture? Suggest a possible explanation if there is a conflict.

7. Calculate the number density of conduction electrons and holes in pure Ge at 300 K (assume $m_n = m_p = m_{electron}$; $E_g = 0.67$ eV).

8. Draw a schematic plot ($\ln\sigma$ versus T^{-1}) of the temperature (T) dependence of the electrical conductivity (σ) of an extrinsic semiconductor; label the different regions and the meaning of the slopes.

9. Slater antiferromagnetic ordering, which causes metal–insulator transition, induces the doubling of the magnetic unit cell. What experimental technique would be appropriate to detect such a transition?

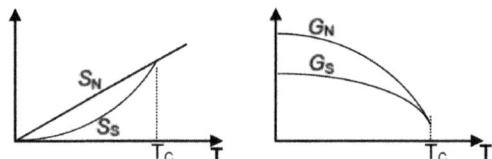

Figure 8.3 Schematic plot of the entropy (S) and free energy (G) with temperature (T) in the superconducting (S) and normal states (N); T_C is the critical temperature (Exercise 8.4.B11).

10. Arrange the materials – stainless steel, silver, diamond – in order of increasing: (i) electrical resistivity and (ii) thermal conductivity. Which of these materials is the best choice for an application that requires high electrical resistivity and high thermal conductivity?

11. Give qualitative explanations for the variation of entropy (S) and free energy (G) with temperature, of the superconducting (S) and normal (N) states, shown schematically in Figure 8.3. Suggest how the data for the normal state could be obtained below the T_C.

12. Name some organic materials that show ferroelectric behavior and note briefly how electric polarization is reversed in these materials, upon reversing the applied electric field.

13. Solid superionic conductors are used in batteries, fuel cells, electrochromic devices, sensors *etc.* Their electrical conductivity depends on temperature as, $\sigma(T) = \dfrac{a}{T} e^{-\frac{b}{T}}$. Explain how this temperature dependence arises and the factors that contribute to a and b.

8.5 Chapter 5

A. Content Based

1. Find the expression relating the effective Bohr magneton number (p) for a paramagnet to the molar paramagnetic susceptibility, χ at temperature T.

2. Which of the following are expected to show a negative magnetic susceptibility: (i) water, (ii) crystalline sodium, (iii) mercury at 1 K and low magnetic field? Explain why.

3. Strike out the inappropriate descriptions in Table 8.1.

4. Draw schematic plots of χ^{-1} *versus* T for paramagnets and ferromagnets (above their Curie temperature), and explain the relative trends.

5. Show that, within the free electron model, the Pauli paramagnetic susceptibility can be written as: $\chi_{\text{Pauli}} = \mathfrak{D}(\varepsilon_F)\beta^2$, where

Table 8.1 Properties and examples of diamagnetic and paramagnetic materials.

		Diamagnetism	Paramagnetism
Magnetic susceptibility	Sign	Positive/negative	Positive/negative
	Magnitude	Small/large	Small/large
	Temperature	Dependent/independent	Dependent/independent
Example		He/CuSO$_4$/superconductor (below T_C)	He/CuSO$_4$/superconductor (below T_C)

Table 8.2 Ratio of the inter-atomic distance (r_{ab}) to 3d-orbital radius (r_o) for some metals.

	Fe	Co	Ni	Cr	Mn	Gd
r_{ab}/r_o	3.26	3.64	3.94	2.60	2.94	3.10

$\mathfrak{D}(\varepsilon_F)$ is the density of electronic states at the Fermi energy and β is the Bohr magneton.

6. What is the saturation magnetic moment (spin-only) of Fe$_3$O$_4$ per formula unit expressed in terms of the Landé factor and Bohr magneton?

B. Context Based

1. Using a dimensional analysis show that the formula for the Langevin diamagnetic susceptibility (molar) leads to the *volume mol^{-1}* unit.

2. Assuming a radius of 0.7 Å, calculate the molar diamagnetic susceptibility of C atom.

3. Draw schematic plots of the $\dfrac{\text{magnetization}}{\text{magnetic field}}\Big/\text{temperature}$ $\left(M \text{ versus } \dfrac{H}{T}\right)$ of paramagnetic ions with different values of the total angular momentum quantum number (consider the full expression for the magnetization involving the Brillouin function).

4. A monoatomic ferromagnet has a Curie temperature of 630 K and each atom contributes 0.6 Bohr magneton towards the magnetic moment. Calculate the internal field in this material.

5. What are the relative advantages and disadvantages of replacing iron cores by ferrite cores in transformers?

6. Why is it desirable to prepare acicular ferromagnetic particles for applications in magnetic recording materials?

7. Given the data in Table 8.2 for the ratio of inter-atomic distance (r_{ab}) to the 3d-orbital radius (r_o) for various metals, explain why

Mn $(r_{ab} = 2.58$ Å$)$ is not ferromagnetic while MnAs $(r_{ab} = 2.85$ Å$)$ and MnSb $(r_{ab} = 2.89$ Å$)$ are ferromagnetic.

8. What are soft and hard ferromagnets? Show the difference in their hysteresis curves in a schematic plot. Identify examples of soft and hard magnetic materials.

9. $[Mn_{12}O_{12}(CH_3COO)_{16}(H_2O)_4]4H_2O \cdot 2CH_3COOH$ is the classic example of a single molecule magnet (SMM). The molecule has an outer ring of eight Mn^{III} ions with a tetrahedral cluster of four Mn^{IV} ions in the interior. The ions in the outer ring and those in the interior interact antiferromagnetically. What is the total spin quantum number, S of this molecule? Draw a schematic diagram of the energy levels corresponding to the magnetic quantum numbers, m_S in the absence and presence of an external magnetic field, and state how the SMM shows magnetic hysteresis.

8.6 Chapter 6

A. Content Based

1. In the discussion on surface plasmon resonance absorption, it is noted that the surface area to volume ratio of spherical particles increases with the radius r, as $\frac{3}{r}$. Show how the surface area to volume ratio varies with the edge length a, in the case of cubic particles.

2. What is an exciton? How does the application of an electric field on a suitable material lead to the creation of excitons?

3. What are the materials properties of silver halides that are relevant for their use in photography? Explain how they are relevant.

4. Give one example of an organic polymer that exhibits electroluminescence. What specific structural features are important that can lead to a practical application?

5. Cite examples of crystals of achiral and chiral molecules that produce optical second harmonic generation.

6. Urea crystals can produce optical second harmonic generation (SHG) effects. What features in the molecular structure and crystal structure are important for this? Why?

B. Context Based

1. Using the free electron model, the charge displacement in a field, $E(\omega)$ is found to be $\frac{eE}{m\omega^2}$, where e = charge, and m = mass of the electron. Using dimensional analysis, verify that this expression shows length unit for the displacement.

2. Expressions for the frequency dependent dielectric constant are, $\sqrt{\epsilon(\omega)} = n(\omega) + ik(\omega)$ and $\varepsilon(\omega) = 1 - \dfrac{\omega_P^2}{\omega^2}$ where ω_P is the plasma frequency, and the relation for reflectivity is, $R(\omega) = \dfrac{(1 - n(\omega))^2 + (k(\omega))^2}{(1 + n(\omega))^2 + (k(\omega))^2}$. Based on these, make a schematic plot of the reflectivity as a function of ω.

3. Poly(N-vinylcarbazole) is a well-studied organic photoconductive polymer. What is its chemical structure and dark conductivity? Why does the addition of molecules like 2,4,7-trinitro-9-fluorenone change its conductivity?

4. The luminescence of semiconductor nanoparticles (quantum dots) is known to shift to lower energies (higher wavelengths) when the particle size increases. For example, when the size of glutathione-capped CdTe nanoparticles increases from ~2.4 nm to ~3.4 nm, the emission peak shifts from ~535 nm to ~610 nm. Explain the basis for this.

5. Draw a schematic energy level diagram for a pyrene based excimer laser and state how it works.

6. Solar cells based on perovskite materials are under intense exploration. Give examples of the materials and note briefly their advantages and disadvantages.

7. Optical limiting is an important nonlinear optical (NLO) application. What is this phenomenon? Cite examples of NLO processes that can cause optical limiting. List examples of organic materials used as optical limiters.

8. What is electrochromism? Give examples of well-known electrochromic materials and the basic principle of their action.

9. What is the twisted nematic effect, and how is it used in liquid crystal based display devices?

8.7   Chapter 7

A. Content Based

1. What are some of the common methods used for the synthesis of polymers?

2. For the process of homogeneous nucleation from the melt, derive an expression for the product, $r^* \cdot |\Delta T|$, where $r^* = $ critical radius for nucleation and ΔT is the supercooling. What properties of the material does this product depend on?

3. Consider the ternary phase diagram for the $\frac{1}{2}Y_2O_3$–BaO–CuO system. Determine the mol% of each component in the compositions, Y_2BaCuO_5 and $YBa_2Cu_3O_{6.5}$.

4. Describe the electrocrystallization method used to prepare crystals of organic conductors and superconductors, citing examples.

5. Represent the basic chemical reaction steps involved in the sol–gel synthesis of silica starting from a tetralkoxysilane precursor.

6. What are the main steps involved in the chemical synthesis of metal nanoparticles through the colloidal route?

B. Context Based

1. The basic components of steel are iron and carbon. Draw a schematic phase diagram of the iron (Fe)–iron carbide (cementite, Fe_3C) system below 1000 °C, and indicate the relevant phases.

2. Martensitic transformation is an important solid state phase transition in steel. How is this phase transition achieved? State the important aspects of the transition.

3. Given the following data for a material, melting point $= 600$ K, enthalpy of fusion $= 2.37 \times 10^5$ kJ m^{-3}, surface tension $= 33 \times 10^{-5}$ N m^{-1}, estimate the critical radius for nucleation of the material from its melt when supercooled by 80°.

4. What are zeolites? What are the main steps involved in their hydrothermal synthesis?

5. What is the Stöber process? How is it adapted to produce mesoporous silica?

6. Electron diffraction experiments show that gold nanoparticles have an fcc lattice structure with a unit cell length of 4.08 Å. Estimate the approximate number of gold atoms in a spherical particle having a diameter of 10 nm.

7. Describe an approach to control the synthesis of semiconductor nanoparticles to achieve color tuning.

8. Polyaniline is one of the most popular conducting polymers. What are the general forms of polyaniline and the method for their synthesis?

9. How does the rate of cooling a melt/liquid lead to the formation of a glass instead of crystal? Give examples of inorganic and organic glasses noting briefly how they are prepared.

10. What are the general methods for the synthesis of fullerenes?

Postface

The formulation and development of this book were guided by the goal of providing insight into the core concepts that form the basis of a course on materials chemistry. The sequential and perhaps terse presentation, aided by graphics support is meant to help the student grasp the key ideas without having to delve into meandering details. As briefly noted in the Preface, the subject of materials chemistry is an expansive and steadily expanding one. Coverage of the vast repertoire of specialized concepts and the innumerable classes of materials is beyond the scope of this book.

Accounts of various families of materials could serve as a useful supplement to the present coverage of the basic themes and principles. However, the classification of established and emerging materials itself is a non-trivial exercise. It could be based on functionality, for example materials of interest in applications like electronics, photonics, spintronics, plasmonics, sensing and actuation, catalysis, energy generation and storage, environmental remediation, biology and medicine (including theranostics). The basis could be the essential materials structure and specific attributes – liquid crystals, glasses and disordered systems, ionic/superionic conductors, porous solids and liquids, polymers including conducting polymers, molecular materials, various classes of nanomaterials, biomaterials *etc.* Another approach would be to focus on advanced materials features as in the case of shape memory materials, topological insulators, multiferroics, giant and colossal magnetoresistance materials, high T_C superconductors, metamaterials, photonic crystals, quantum materials, smart materials of various hues, phase change materials and so on. Indeed there is significant overlap between several of the

Core Concepts for a Course on Materials Chemistry
By T. P. Radhakrishnan
© T. P. Radhakrishnan 2023
Published by the Royal Society of Chemistry, www.rsc.org

groups listed. These considerations highlight the complexity involved in developing a comprehensive coverage of types and classes of materials. In a typical course, it would therefore be prudent to focus on selected topics of interest in the emerging scenario and context, using relevant and state-of-the-art review articles, or perhaps a supplementary and exhaustive textbook that may eventually evolve.

Even though the main text itself does not cover specialized materials for the reasons discussed above, we have attempted to provide a glimpse into some of the specific classes and families of materials through the Exercises; these include liquid crystals, ionic conductors, single molecule magnets, solar cell materials, conducting polymers, glasses, zeolites and mesoporous materials.

T. P. Radhakrishnan
Hyderabad

Further Reading

1. H. V. Keer, *Principles of the Solid State*, Wiley-Interscience, 1993.
2. E. A. Moore and L. E. Smart, *Solid State Chemistry: An Introduction*, CRC Press, 2021.
3. M. T. Weller, *Inorganic Materials Chemistry*, Oxford University Press, 1995.
4. *Nanoscale Materials in Chemistry*, ed. K. J. Klabunde, Wiley-Interscience, 2001.
5. *Materials Science and Engineering, An Introduction*, ed. W. D. Callister and D. G. Rethwisch, Wiley, 2018.
6. C. Kittel, *Introduction to Solid State Physics*, Wiley, 2005.
7. B. D. Fahlman, *Materials Chemistry*, Springer, 2011.
8. H. R. Alcock, *Introduction to Materials Chemistry*, Wiley, 2019.
9. Web resource: http://chemistry.uohyd.ac.in/~CY551/

Core Concepts for a Course on Materials Chemistry
By T. P. Radhakrishnan
© T. P. Radhakrishnan 2023
Published by the Royal Society of Chemistry, www.rsc.org

Subject Index

absorption (optical/light) 145–146
    electro-optic effects and 163
    intrinsic semiconductors and 85
    lattice defects and 46
ac Josephson effect 104
acid–base reactions 178
acoustic phonons 51, 52, 58, 82, 147
adiabatic demagnetization 121–123
aliovalent impurity 40
alkali metal halides, color centers 45
alkanethiol assembly on crystalline gold
    surface 219
alkyl silane adsorption on glass 219
alloys, superconducting 103
alternating current (ac) Josephson
    effect 104
alumina (Al$_2$O$_3$)
    nanoporous alumina membrane
        synthesis 216
    ruby laser 161
    sintering 202
aluminium thin films 204
amphiphiles 221–222, 223–224
anode of photoelectrochemical cell 166
anodization 207
    nanoporous alumina membrane
        synthesis 216
antiferromagnetism 113, 133–134,
    135, 136
anti-Stokes lines 147
anti-Stokes shifted luminescence 155
atom
    free 72, 73, 117
    orbitals see orbitals
atomic scattering factor 22, 29, 30
Au see gold

ball-milling 212
bands see energy bands
Bardeen, Cooper and Schrieffer (BCS)
    superconductors 100–101, 102
basis (lattice) 11
    multi-atom 52
    spherical 9, 11, 13
BCS (Bardeen, Cooper and Schrieffer)
    superconductors 100–101, 102
BEDT-TTF (bis(ethylenedithio)-
    tetrathiafulvalene) 102
beta-quartz 195
Bethe–Slater curve 125

binding energy (of electrons)
    Cooper pairs 100
    in electron spectroscopy 148
    exciton 145, 149
biocrystallization 185
bioluminescence 159
bipolar junction transistor (BJT) 90
bis(ethylenedithio)tetrathiafulvalene
    (BEDT-TTF) 102
Bloch functions 78–80
Bloch wall 131
blocking temperature 137
body-centered lattices 8, 21
    cubic (bcc) 9, 20, 23, 27, 28, 29, 44
Boltzmann distribution and
    paramagnetism 117
boride, superconducting 103
Bose–Einstein distribution 54, 66
bottom-up (build-up) fabrication of
    nanomaterials 211, 213–215
    ultrathin films 219
Bragg's law for diffraction 16–17, 21,
    23, 29, 32
Bragg's reflection 76, 77, 81
Bravais lattice 6–10, 13, 19, 20, 24,
    25, 26, 27
break-down fabrication see top-down
    fabrication
Bridgman–Stockbarger method 198–199
Brillouin function 118–119
Brillouin scattering 147
Brillouin zones 51, 52, 76–78, 80, 81,
    99, 145
bromide salt of tetrathiafulvalene 45
build-up fabrication see bottom-up
    fabrication
Burger's vector 35, 47

calcination 181, 202, 215
carbon
    nanotubes 216
    superconductors based on 103
cathode of photoelectrochemical cell 167
cavity, resonating 161–162, 163
centered rectangular lattice 6
centrosymmetric media/materials 170, 172
ceramic materials, crystals 200
ceramic method 181
charge compensation defect 2
charging, xerography 150

chemical potential 66, 185
chemical vapor deposition (CVD)
    206–207, 215, 216, 224–225
chemiluminescence 158–159
chirality, non-centrosymmetric crystal
    lattice 172
chlorosilane adsorption on glass 219
chromium (Cr^{3+}) ions, ruby laser 161
cinnamic acid, solid-state dimerization
    182–183
close packing, cubic and hexagonal 10
coating (thin film deposition)
    photolithography 208
    spin-coating 203–204, 218
    cobalt (ii) oxide (CoO) 94
coherence length 101
colloids 208–210, 213, 214
complex dielectric function 142–143
composites
    nano-sized (nanocomposites) 210,
        217–219
    negative thermal expansion 61
conduction (electrical) and conductivity
    68–70, 82
    conduction band edge 84
    light increasing
        (=photoconduction) 148–152
    Ohm's law and 68–70
    semiconductors
        extrinsic 88–90
        hopping 94
        intrinsic 87–88
conduction (thermal) 62–63
    metals 63, 71–72
contact angle and nucleation 189
cooling, crystal growth by 193–195
Cooper pairs 100–101, 102
cooperative phenomena in magnetism
    113, 123–137
copper (Cu) metal
    Mn introduced into 39
    non-ferromagnetism 128
core–shell nanostructures 215–216
corrosion 180–181
covalent solids 1–2
critical current with superconductors 07
critical field, superconductors 97
critical radius and nucleation 187, 188, 189
critical temperature, superconductors
    96, 99, 101
crystal(s) 2, 73–75, 184–202
    defects in *see* defects
    energy spectrum of electrons in
        73–75

growth *see* crystallization
lattice *see* lattice
non-linear optic applications
    171–172
    second harmonic generation 174
structural motifs 33
symmetry *see* symmetry
thermal properties *see* thermal
    properties
crystal-to-crystal transformation 184
crystallization (and crystal growth)
    184–202
    from melts 185–186, 198–199
    process of and application
        184–185
    from solution 190–197
cubic close packing (ccp) 10
cubic lattice 8, 13
    body-centered (bcc) 9, 20, 23, 27,
        28, 29, 44
    face-centered (fcc) 9, 20, 27
Curie law 116, 119
Curie temperature 126, 128, 137
Curie–Weiss law 127, 134
current with superconductors 97, 100,
    103–104
    critical 07
    persistent 97
Czochralski (crystal pulling) method 199

dark conductivity 149
dc Josephson effect 103
Debye model of heat capacity 56–59
Debye temperature 82, 101
defects (crystal) 35–40
    characterization 45–47
    extrinsic 35–36, 39–40
    intrinsic 35–36, 36–39
definite proportions, law of 40
demagnetization
    adiabatic 121–123
    cycles of magnetization and
        130, 133
density of states
    Debye model for heat capacity
        56–57
    electronic 67–68
derivatization and crystallization
    195–197
development
    photography 152
    photolithography 208
    xerography 150
devitrification 199

diamagnetism 111–116, 112, 113–116
    example materials 112
    Landau 121
    superconductors 97
    susceptibility to 112, 115–116
diamond 33, 83, 88–89
dielectric constant, light and 143–145
dielectric function, complex 142–143
dielectric polarization (dielectrics)
    104–110, 143–144
dielectric susceptibility (electric
    susceptibility) 105, 143, 168
diffraction
    electron 31–33
    neutron 31
    X-ray *see* X-ray diffraction
diffusion
    crystal growth by 193, 194, 196,
        201
    in solid states 179–180
dimensions (solids) 2
    defects and 35
    see also specific dimensions
dimerization of cinnamic acid 182–184
diodes
    laser 162–163
    light-emitting (LEDs) 157, 163
dip-coating 203
direct band gap semiconductors 85,
    154, 163
direct current (dc) Josephson effect 103
displacement and the dielectric constant
    143–144
disproportionation (thin film
    deposition) 206
DNA origami 215
domains (magnetic), ferromagnetic ma-
    terials 129, 130–133
doped semiconductors 39, 88–89
down-conversion 154
drift velocity 69
drop-casting 203
Dulong–Petit law 55, 58
dye lasers 161
dye-sensitized solar cells 165

edge dislocation 35
effective mass 80–81, 86, 88
Einstein's model of heat capacity
    55–56, 58
electric susceptibility (dielectric
    susceptibility) 105, 143, 168
electrical energy from photovoltaic
    cells 164

electrical properties 64–110, *see also*
    conduction; current; dielectric
    polarization; resistivity; thermo-
    electric effects; voltage
electrochemical thin film deposition 207
electrocrystallization 196–197
electrodeposition 215
electrolithography, nanomaterials 213
electroluminescence 156–157, 164
electromagnetism
    in electron diffraction 33
    Faraday's law of electromagnetic
        induction 140
    Lenz's law 113
electron(s)
    density 30, 31
        of states 67–68
    energy *vs.* wave vector 76–78, *see*
        also energy bands
    excitation 85, 86, 87
        light and 153, 154, 155
    Fermi sea of 99
    free *see* free electrons
    magnetism and 111, 123–125, 141
        angular momentum 111,
            114, 117, 118, 119
        antiferromagnetism 134
        diamagnetism 113–115
        ferromagnetism 125,
            128, 129
        paramagnetism 116–121
        spin 111, 117, 120–125, 126,
            128–129, 130, 131,
            133–135, 141
    missing 85, 86
    nanomaterials and
        confinement of 211
    orbital *see* orbitals
    pairs (Cooper pairs) in super-
        conductors 100–101, 102
electron beams 31–32
    nanomaterials 213
    thin film deposition 205, 206,
        207, 208
electron diffraction 31–33
electron spectroscopy 148
electro-optic effects 173–175
electrophotography (xerography)
    150–151
electroplating 207
elements, superconducting 96, 103
energy
    binding *see* binding energy
    electron *see* electrons

energy (*continued*)
    exchange *see* exchange energy
    Fermi 65, 66, 68, 72, 81, 99,
       101, 120
    free *see* free energy
    potential *see* potential energy
energy bands 72–81
    gap (energy gap; E_g) 76–78, 83,
       84–85, 87, 88
       light and 149, 154, 163, 166
    paramagnetism and 120
energy gap
    in band structure *see* energy bands
    superconductors 98–100
epitaxy, metal organic vapor phase 207
ethylene 74, 78, 79
eutectic point 191
evaporation 203
    crystallization by 192
    thermal 204–205
evaporation/condensation crystallization
    193, 201
Ewald construction/sphere 28–29,
    32–33
exchange energy 123–125
    antiferromagnetism and 133
    ferromagnetism and 125
excitons 145–146, 154, 156, 164
exfoliation (2-D material formation) 225
exhaustion range 90
expansion, thermal 59–61
exposure
    photography 152
    photolithography 208
    xerography 150

F-centers 45
fabrication/synthesis 177–225
face-centered lattices 8, 21
    cubic (fcc) 9, 20, 27
Faraday gold colloids 214
Faraday law of electromagnetic
    induction, electromagnetism 140
Faraday method for determining
    magnetic susceptibility 137, 139–140
Farbe (F)-centers 45
fatty acid monolayers 219, 221–222
Fermi–Dirac distribution 66, 68, 120
Fermi energy 65, 66, 68, 72, 81, 99,
    101, 120
Fermi gas, free electron 64–65
Fermi level 68, 75, 88, 100, 163, 166
Fermi sea of electrons 99
Fermi sphere 67, 68, 69

Fermi temperature 72, 120
fermions 66, 123
ferrimagnetism 113, 134–136, 137
ferrites 135
    hopping semiconductors based
      on 94
ferroelectric materials 109–110
ferromagnetism 113, 125–129, 130, 131,
    133, 134, 136, 137
    magnetization 126–128, 129,
      130, 132
ferrous oxide (FeO) 41, 94
films
    thin 202–208
    ultrathin (nanosheets) 210, 213,
      219–225
fixing
    photography 151
    xerography 150
flash evaporation 205
fluorescence 15–16, 153
    nanoclusters 211
    special mechanisms 154–156
fluorite 33, 44, 158
Frätzel cell 165
free atom 72, 73, 117
free electrons (and free electron theory/
    model) 64–72, 76, 77, 80, 81, 82,
    119
    Schrödinger equation 65, 73, 78
free energy
    crystallization 186–188, 190, 201
    metal oxide formation 181
Frenkel defects 39
Frenkel excitons 145–146

gas, free electron, heat capacity 68
gaseous state, crystal growth in 185
gel, cooling in a (for crystal growth) 194,
    see also sol–gel method
germanium (Ge) lattice 88–89
glass, chlorosilane or alkylsilane
    adsorption on 219
glide reflection (crystals) 3
gold (Au)
    alkanethiol assembly on crystal-
      line gold surface 219
    nanoparticles 216
      Faraday's 214
    resistivity 82
Gouy method for determining magnetic
    susceptibility 137, 138–139
graphene 224–225
grinding 202, 211–212

Hall effect (and Hall field and coefficient) 70–71
Hamiltonian 124, 133
harmonic generation
    second (SHG) 169–171, 171–172
    third (THG) 175–176
heat
    conduction 62–63
    expansion with 59–61
heat capacity (lattice) 54–59
    free electron gas 68
    superconductors and 99–100
Heisenberg Hamiltonian 124, 133
Helmholtz free energy 186, 187–188
heterogeneous nucleation 188–189
hexagonal close packing (hcp) 10
hexagonal lattice/systems 5, 6, 8, 10, 107
high-temperature superconductors 44
holes (in orbitals) 85–86, 87
    electron–hole pairs (=excitons) 145–146, 154, 156, 164
homogeneous nucleation 186–188, 189
honeycomb lattice 10, 224
hopping semiconductors 93–95
Hückel approach 74
Hund's rule 125
hydrothermal crystal growth 195
hysteresis (magnetic) 129–130

impurities
    aliovalent 40
    hopping semiconductors and doping of impurity metal ions 94
    sintering 202
    substitutional/interstitial 40–41
    zone-refining 200
incandescence 158
indexing in powder X-ray diffraction 17–18
indirect band cap semiconductors 85, 154
infra-red irradiation and photoconduction 149
injection laser 162–163
inorganic materials
    for second harmonic generation 171
    synthesis and fabrication 177–178
insulators
    semiconductors and 83, 95, 99
    superconductor–insulator–super-conductor junction 103

intercalation method (2-D nanosheets) 225
interferometer, Mach–Zander 174
interstitial impurity 40–41
inverted population (population inversion) 160–161, 162
ion beam sputtering 206
ionic solids 1
iron (II) oxide (ferrous oxide) 41, 94
iron garnets 136
isothermal crystallization 192–193
isotope effect, superconductors 101

Jahn–Teller effect 95
Josephson effects and devices 102–104

Kerr effect 174–175, 175
Kerr gate 175
Knudsen cell 197

laminar flow system, crystal growth by cooling in 194–195
Landau diamagnetism 121
Langmuir–Blodgett film 221–224
Larmor (precession) frequency 113–114, 115
laser(s) 159–163
    Nd:YAG 161, 171, 205
    thin film deposition 205
laser ablation 205, 213
laser diode 162–163
lattices (crystal) 3, 11–13
    body-centered *see* body-centered lattices
    Bravais 6–10, 13, 19, 20, 24, 25, 26, 27
    cubic *see* cubic lattice
    face-centered *see* face-centered lattices
    hexagonal 5, 6, 8, 10, 10, 107
    honeycomb 10, 224
    monoclinic 6, 8, 107
    orthorhombic 8, 9, 107
    reciprocal 24–28, 29–30, 33, 53, 57, 76
    tetragonal 6, 8, 9, 20, 107
    thermal properties 48–63
    triclinic 20, 26, 107
    trigonal 6, 8, 107
    vibrations *see* vibrations
layer-by-layer assembly 220–221
LEDs (light emitting diodes) 157, 163
Lenz's law in electromagnetism 113

light *see* optical properties *and entries under* photo-
light-emitting diodes (LEDs) 157, 163
line defects 35, 45–46
linear polarization response 167–169
liquid (state), crystal growth from 185,
  see also vapor–liquid–solid method
liquid-crystals 2
liquidus 191
lithium niobate (LiNbO$_3$)
  electro-optics 171
  piezoelectrics 161
lithography
  nanomaterials 212–213
  thin films 207–208
localized surface plasmon resonance
  (LSPR) extinction 211
London penetration depth 101
Lorentz force 68, 70, 114, 143
luciferin 159
luminescence 152–159
luminol 158–159

M-centers 45
Mach–Zander interferometer 174
macroscopic quantum interference
  104
magnet, permanent 133
magnetic field, definition 140
magnetic flux 101, 104, 113, 140
  density (=magnetic induction) 141
magnetism 111–141
  basic definitions 140–141
  cooperative phenomena 113,
    123–137
  hysteresis 129–130
  spontaneous 126, 127, 129, 141
  superconductors 97, 98, 99, 101,
    102, 104
magnetite (Fe$_3$O$_4$) 135–136
magnetization 140–141
  cycles of demagnetization and
    130, 133
  definition 137, 140–141
  saturation 128
  susceptibility to 137–140, 141
    antiferromagnetism and 134
    diamagnetism and 112,
      115–116
    experimental determination
      137–140
    ferrimagnetism and 135–136
    ferromagnetism and
      126–128, 129, 130, 132

paramagnetism and 112,
  116, 119–121
superparamagnetism and
  136, 137
zero 130, 131, 132
magnetocrystalline anisotropy 130,
  131, 136
magnetostriction 130, 131
magnetron sputtering 206
manganese (Mn) introduced into copper
  (Cu) metal 39
manganese oxide (MnO)
  antiferromagnetism 133–134
  hopping semiconductors 94
mass, effective 80–81, 86, 88
mass transport in solid state, sintering
  and 201
material synthesis/fabrication 177–225
Matthiessen's rule 70
mean field approximation 126–127, 134
mechanoluminescence 158
Meissner effect 97
melts of solids, crystallization from
  185–186, 198–199
metal(s) 81–95
  corrosion 180–181
  electrical resistivity and
    conductivity 82
  nanoparticles and nanostructures
    218, 221
    noble metals 214
  plasma frequency 145
  semiconductors and 81–95
  solid 2
  thermal conductivity 63, 71–72
  thin films 204
  transition *see* transition metals
metal complexes
  formation 178
  optical properties 154
metal ions, impurity, hopping semiconductors and doping of 94
metal organic chemical vapor deposition
  206–207
metal oxides (MOs)
  anodized 207
  from corresponding metal
    ores 181
  nanoparticles 214
  non-stoichiometry 41–43
  thin films 205
micelles, reverse 214
micro-emulsion technique 214
Mie scattering 147

Miller planes and lines 13–14, 17, 19, 20, 21, 22, 23, 26, 28, 32
mobility (electron) 70, 82
molecular orbitals *see* orbitals
molecular solids 2
monoclinic crystal system 6, 8, 107
monolayers 219–223
 layer-by-layer 220–221
 self-assembled 219
multi-atom basis (lattice) 52
MXenes 224–225

n-type semiconductors 39, 89, 90
 junctions between p-type
  see p–n junctions
 light and 165, 166
nanoclusters 211
nanocomposites 210, 217–219
nanomaterials 210–219
 sizes 210, 211
 specialized examples 215–219
nanoparticles, typical sizes 210
nanorods and nanowires 210, 216–217
nanosheets (ultrathin films) 210, 213, 219–225
nanostructured materials (nanostructures) 210, 211, 213, 215–219
 specialized examples 215–219
nanowires and nanorods 210, 216–217
Nd:YAG laser 161, 171, 205
Néel relaxation time 136
Néel temperature 134
negative thermal expansion 60–61
neodymium (Nd):YAG laser 161, 171, 205
Neumann's principle 105–106
neutron diffraction 31
nickel
 ferromagnetism 128–129, 141
 spontaneous magnetization 129, 141
nickel oxide (NiO) 94
nickel telluride (NiTe) 43
noble metal nanostructures 214
non-centrosymmetric structure
 point groups 106, 108
 for quadratic non-linear optics 172–173
non-linear optics, basic principles 167–176
non-solvent, addition or diffusion of (in crystal growth) 193
non-stoichiometry 40–45
nuclear spin, magnetic moment and 111

nucleation 185, 186–190
 classical theory 186, 188, 189, 190
 heterogeneous 188–189
 homogeneous 186–188, 189
 two-step 189–190
nucleic acid-based nanostructures 215

Ohm's law and electrical conduction 68–70
OLEDs (organic LEDs) 157, 163
one-dimensional (1-D) crystal lattice
 vibrations 48–49, 49–50
 X-ray diffraction 21–22
Onnes, Kamerlingh 96
optical phonons 52, 58, 147
optical properties 142–176, *see also entries under* photo-
 absorption *see* absorption
 non-linear optics, basic principles 167–176
orbitals (electron/molecular/atomic) 67, 72–75, 78, 79, 80
 angular momentum, magnetism and 111, 114, 117, 118, 119
 free atom 72
 vacant *see* holes
order in solids 2
organic LEDs (OLEDs) 157, 163
organic molecules/compounds/materials
 fluorescence 154
 photovoltaics based on 165
 polymeric, third harmonic generation 175–176
 synthesis and fabrication 177–178
organic solids 44–45
organic superconductors 103, 197
organometallic chemical vapor deposition 206–207
orthorhombic systems and lattices 6, 8, 9, 107
oxides
 of metals *see* metal oxides
 sol–gel synthesis of oxide nanoparticles 214–215
 superconducting 103
oxygen and corrosion 180–181

p–n junctions 90, 93
 diode laser 163
 photovoltaic cells 164
p-type semiconductors 39, 89, 90
 junctions between n-type and
  see p–n junctions
 light and 166, 167

paramagnetism 111–112, 116–123
    example materials 112
    Pauli 119–121
    quantum theory 117–119
    susceptibility to 112, 119–121
Pauli paramagnetism 119–121
Pauli principle 65
Peierls instability/distortion 80, 95
Peltier effect 92–93
penetration depth, London 101
Penrose tiling 34
permutation symmetry 123
perovskite 33, 44, 110, 165
persistent current with
    superconductors 97
phase diagram
    crystal growth from solution
        191–192
    solid-state reactions 181, 182
phonons and their dispersion 48–54,
    54–55, 58, 60, 62, 63, 67, 82, 85, 157
    acoustic phonons 51, 52, 58,
        82, 147
    optical phonons 52, 58, 147
phosphorescence 153–154
photochemical solid-state reactions 182
photoconduction 148–152
photoelectrochemical cell 165–167
photoelectrons 148
photoirradiation in photochemical
    cells 166
photolithography
    nanomaterials 213
    thin films 207–208
photoluminescence 153–154
photons 142, 145–148, 154
    superconductors and 99
photopolymerization 184
photoresistor 150
photovoltaic cells 164–165
physical vapor deposition (PVD) 197, 204
piezoelectric materials 105–108, 110
plane defects 35, 45–46
plasma-based deposition 205
plasma frequency 144, 145
plasmon 145
    frequency 145
plasmon resonance
    localized surface plasmon reson-
        ance (LSPR) extinction 211
    surface plasmon resonance
        absorption 211
pnictide superconductors 103
Pockel's effect 174

point groups (of crystals) 106
    non-centrosymmetric 106, 108
    piezoelectric materials 105,
        106, 108
    pyroelectric materials 108
    symmetry 3, 4, 5, 6, 8, 13
polariton scattering 147
polarization
    dielectric polarization 104–110,
        143–144
    optical polarization response,
        linear and non-linear 167–169
poling 172–173
poly(acetylene) 74, 75, 78, 81
polyanions in layer-by-layer assembly 220
polycations in layer-by-layer
    assembly 220
poly(diacetylene) 176, 184
polymer(s)
    organic, third harmonic
        generation 175–176
    superconductor 103
polymerization
    light-induced
        (photopolymerization) 184
    thin film deposition 206
polymorphism 185
polyol method 214
poly(phenylene vinylene) (PPV) 165,
    176
poly(vinyl alcohol) (PVA) thin film, silver
    nanoparticles in 218
poly(vinylidine fluoride) 110
population inversion inverted popu-
    lation 160–161, 162
potassium dihydrogen phosphate,
    KH_2PO_4 (KDP) 171
potential energy
    free electron theory and 64, 73
    thermal properties and 50, 59, 60
powder X-ray diffraction 17–20
praseodymium oxide 43–44
precession frequency, Lamor's
    113–114, 115
primitive vectors (crystals) 8–9, 24, 26
printing (positive image) in
    photography 152
pulsed laser deposition (PLD) 205
pyroelectric materials 108–109, 110

quadratic non-linear optic effects 170,
    171–173, 174–175
quantum dots 210
quantum interference, macroscopic 104

quantum theory of paramagnetism
    117–119
quartz crystal microbalance 203
β-quartz 195
quasi-crystals 33–34

R-centers 45
radiofrequency sputtering 206
radius (critical) and nucleation 187,
    188, 189
Raman scattering 147
Rayleigh scattering 147
reactive evaporation 205
reciprocal lattice 24–28, 29–30, 33,
    53, 57, 76
rectangular lattice 6
rectification
    optical 170
    p–n junction 90
reflectance and reflectivity 142–143
reflection (crystals) 3
    Bragg 76, 77, 81
    glide 3
refractive index 142, 167, 174, 174–175
relaxation time (magnetization) 136
resistivity (electrical) 70, 83, 96
    metals 82
    zero, superconductors 97, 101,
        102, 102
resonator, optical (resonating cavity)
    161–162, 163
reverse micelles 214
RF sputtering 206
roasting of metal oxides 181
rotation (crystals) 3
ruby laser 161

saturation magnetization 128
scattering
    electron diffraction 33
    light 146–147
        colloids and nanomaterials
            209–210
    neutron diffraction 31
    X-ray diffraction 17, 21, 22, 23, 24,
        28, 29, 30
Schottky defects 38, 47
Schrödinger equation for free electrons
    65, 73, 78
screw dislocation 35
screw rotation (crystals) 3
Seebeck effect and coefficient 91, 92
self-assembled monolayers 219
self-interstitials 38

semiconductors 74, 81–95
    doped 39, 88–89
    extrinsic 39, 83, 88–90, 148, 149
    hopping 93–95
    intrinsic 83–85, 148, 149
    n- and p-type see n-type
        semiconductors; p-type
        semiconductors
    nanostructures 210, 215, 216
    optical properties 145, 146, 148, 150,
        154, 156–157, 159, 163, 165, 166
semiconductor–metal transition 95
    thin films 205, 207
semimetals 74
shake-and-bake method 181
silicates
    formation 180
    negative thermal expansion 60
silicon (Si)
    semiconductor-doped 39
    solar cells based on 165
silver chloride (AgCl) preparation 177–178
silver halides
    Frenkel defects 39
    in photography 151–152
silver nanoparticles in poly(vinyl alcohol)
    (PVA) thin film 218
silver sulfide (Ag$_2$S) formation, Wagner's
    experiment on 179–180
sintering 200–202
sodium (Na), energy spectrum of electrons 4
sodium chloride (NaCl) 178
    color centers 45
    powder X-ray diffraction 17, 19
sodium tungsten bronze (Na$_x$WO$_3$) 44
sol–gel method
    crystallization 200
    oxide nanoparticles 214–215
solar cells 165
solid(s) 1–34, see also vapor–liquid–solid
    method
    crystallization from melts of
        185–186, 198–199
    dimensions see dimensions
    order 2
    organic 44–45
    types/classification 1–2
solid state 1–34
    crystal growth from 197, 199–200,
        201
    special instances of reactions in
        178, 180–184, 182
    structure 1–34
solidus 191

solutions
    crystal growth from 190–197
    evaporation *see* evaporation
solvothermal crystallization 195
spatial order in solids 2
spectroscopy
    electron 148
    Raman 147
sphere
    Ewald 28–29, 32–33
    Fermi 67, 68, 69
spherical basis 9, 11, 13
spin-coating 203–204, 218
spinels 33, 135, 180
spontaneous light emission
    lasers 159, 160
    LEDs 164
sputtering 205–206
SQUID (superconducting quantum
    interference device) 104, 137
stearic acid monolayer 221–222
stimulated light emission
    lasers 152, 159, 160
    LEDs 163
stoichiometry 40
Stokes lines 147
Stokes shift 153, 154
structure factor 21–23, 30
    reciprocal lattice vector and 29–30
sublimation crystal growth 197
substitutional impurity 40–41
sulfides, superconducting 103
superconducting quantum interference
    device (SQUID) 104, 137
superconductors (and superconductivity)
    96–104
    characteristic features 96–98
    discovery 96
    high temperature 44
    materials 102
    organic 103, 197
    theoretical concepts 100–101
    type I 97–98, 101
    type II 98, 101
supercooling and crystallization 188,
    189, 191, 194, 199
superexchange interaction 134
superparamagnetism 136–137
surface free energy and crystallization
    186, 187, 201
surface plasmon resonance
    absorption 211
surface plasmon resonance extinction,
    localized (LSPR) 211

surfactants
    Langmuir–Blodgett films and
        221–222
    reverse-micelles and 214
    self-assembled monolayers and 219
symmetry (crystals) 3–14
    hierarchies 8, 9, 13
    piezoelectric materials 105–108
    point group 3, 4, 5, 6, 8, 13
    translational 3, 3–4, 6, 8, 48
    unit cell 11
synthesis/fabrication 177–225
systematic absence 20–21

temperature, *see also* cooling; heat; heat
    capacity; hydrothermal crystal
    growth; isothermal crystallization *and
    entries under* therm-
    blocking 137
    Curie 126, 128, 137
    electrical conductivity of metals
        and 82
    expansion with 59–61
    Fermi 72, 120
    Fermi–Dirac distribution and 66
    high temperature superconductors
        44
    magnetism
        antiferromagnetism 134, 136
        diamagnetism 112, 116
        ferromagnetism 126, 127,
            128, 129
        paramagnetism 112, 116,
            117, 119, 120, 121,
            122, 123
        superparamagnetism and
            136, 137
    Néel 134
    semiconductors and
        extrinsic 89–90
        intrinsic 83, 85, 87–88
    superconductors and 97, 102
        critical temperature 96, 99, 101
    thermal conduction in high *vs.* low
        temperatures 63
    vacancy and 37
tetragonal systems/lattices 6, 8, 9, 20, 107
tetramethyltetraselenafulvalene
    (TMTSF) 102
tetrathiafulvalene, bromide salt 45
thermal evaporation 204–205
thermal properties 48–63, *see also*
    temperature
    conductivity *see* conduction

thermally-activated delayed fluorescence (TADF) 156
thermocouple 91, 92
thermodynamics of crystallization 185–186, 190
thermoelectric effects 91–93
thermoluminescence 157–158
thin films 202–208
three-dimensions (3-D), crystal systems/ lattices in 5–6
    Bravais lattice 7–10
    multi-atom basis and vibrations 52
    reciprocal lattice 26–27
    Schrödinger equation 65–66
    X-ray diffraction 22
titanium oxide (TiO) 43, 75
TMTSF (tetramethyltetraselenafulvalene) 102
top-down (break-down) fabrication of nanomaterials 211–213
    ultrathin films 219
topotactic reaction 184
transfer (xerography) 150
transition metals
    monoxides, metal/semiconductor behavior 93–94
    non-stoichiometry 40–41
translational symmetry 3, 3–4, 6, 8, 48
transmission and transmittivity (light) 142, 144, 145, 169, 203
transverse acoustic modes 52
triboluminescence 158
triclinic systems and lattices 20, 26, 107
trigonal crystal 6, 8, 107
triplet–triplet annihilation 156
tungsten bronzes 44, 95
two-dimensional (2-D) materials 224–225
    crystal systems in 4–5
        Bravais lattice 6–7, 25–26
        reciprocal lattice 25–26
two-source evaporation 205

ultrathin films (nanosheets) 210, 213, 219–225
ultraviolet photoelectron spectroscopy 148
unit cell symmetry 11
up-conversion 155
urea crystal 172
UV photoelectron spectroscopy 148
vacancy defect 37
valence band 74, 148, 154, 156
    edge 84
    metals and semiconductors and 81, 84, 85, 87, 88, 89, 90

vanadium (v), non-stoichiometry 41
vanadium (ii) oxide (VO) 94
vanadium (iii) oxide (V_2O_3) 95
vapor(s)
    crystal growth from 197
    deposition 197, 203, 204–205
        chemical (CVD) 206–207, 215, 216, 224–225
        physical (PVD) 197, 204
vapor–liquid–solid method (nanowire formation) 216
vibrating sample magnetometer 140
vibrations, lattice 48–54
    quantization 54–55
voltage, temperature gradient across material causing the generation of 91–93
volume free energy and crystallization 186, 187
von Laue condition 23–25
vortex state 98

Wagner's experiment on Ag_2S formation 179–180
Wannier excitons 146
water, negative thermal expansion 60
wave (and wave vectors)
    electrical properties 69, 76, 80, 85
    interference in X-ray diffraction 15–16
    thermal properties and 49, 51, 53, 54, 57
Wiedemann–Franz law 72
Wigner–Seitz cell 77, 110

X-ray diffraction 15–30, 46
    basic concepts of structure solution and refinement 30
    powder 17–20
X-ray photoelectron spectroscopy 148
xerography and xeroxing 150–151

yttrium barium copper oxide ($YBa_2Cu_3O_7$) 44
yttrium–iron garnet 136

zero magnetization 130, 131, 132
zero resistivity, superconductors 97, 101, 102, 102
zinc oxide (ZnO) 40, 43, 202
zirconium tungstate, negative thermal expansion 60
zone-refining 200, 200